D1557004

PHYSIOLOGICAL ADAPTATIONS

DESERT AND MOUNTAIN

ENVIRONMENTAL SCIENCES

An Interdisciplinary Monograph Series

ARTHUR C. STERN, editor, AIR POLLUTION, Second Edition.
In three volumes, 1968

L. FISHBEIN, W. G. FLAMM, and H. L. FALK, CHEMICAL MUTAGENS: Environmental Effects on Biological Systems, 1970

DOUGLAS H. K. LEE and DAVID MINARD, editors, PHYSIOLOGY, ENVIRONMENT, AND MAN, 1970

KARL D. KRYTER, THE EFFECTS OF NOISE ON MAN, 1970

R. E. MUNN, BIOMETEOROLOGICAL METHODS, 1970

M. M. KEY, L. E. KERR, and M. BUNDY, PULMONARY REACTIONS TO COAL DUST: "A Review of U. S. Experience," 1971

DOUGLAS H. K. LEE, editor, METALLIC CONTAMINANTS AND HUMAN HEALTH, 1972

DOUGLAS H. K. LEE, editor, ENVIRONMENTAL FACTORS IN RESPIRATORY DISEASE, 1972

H. ELDON SUTTON and MAUREEN I. HARRIS, editors, MUTAGENIC EFFECTS OF ENVIRONMENTAL CONTAMINANTS, 1972

RAY T. OGLESBY, CLARENCE A. CARLSON, and JAMES A. MCCANN, editors, RIVER ECOLOGY AND MAN, 1972

LESTER V. CRALLEY, LEWIS T. CRALLEY, GEORGE D. CLAYTON, and JOHN A. JURGIEL, editors, INDUSTRIAL ENVIRONMENTAL HEALTH: The Worker and the Community, 1972

MOHAMED K. YOUSEF, STEVEN M. HORVATH, and ROBERT W. BULLARD, PHYSIOLOGICAL ADAPTATIONS: Desert and Mountain, 1972

In preparation

DOUGLAS H. K. LEE and PAUL KOTIN, editors, MULTIPLE FACTORS IN THE CAUSATION OF ENVIRONMENTALLY INDUCED DISEASE

MERRIL EISENBUD, ENVIRONMENTAL RADIOACTIVITY, Second Edition

PHYSIOLOGICAL ADAPTATIONS

DESERT AND MOUNTAIN

Edited by

MOHAMED K. YOUSEF

DEPARTMENT OF BIOLOGICAL SCIENCES
UNIVERSITY OF NEVADA, LAS VEGAS

STEVEN M. HORVATH

INSTITUTE OF ENVIRONMENTAL STRESS
UNIVERSITY OF CALIFORNIA,
SANTA BARBARA

ROBERT W. BULLARD

DEPARTMENT OF ANATOMY AND PHYSIOLOGY
INDIANA UNIVERSITY, BLOOMINGTON

1972

Academic Press New York and London

ACADEMIC PRESS, INC.
111 Fifth Avenue, New York, New York 10003

United Kingdom Edition published by
ACADEMIC PRESS, INC. (LONDON) LTD.
24/28 Oval Road, London NW1

LIBRARY OF CONGRESS CATALOG CARD NUMBER: 70-187240

PRINTED IN THE UNITED STATES OF AMERICA

Contents

List of Contributors

Numbers in parentheses indicate the pages on which the authors' contributions may be found.

E. F. Adolph, *Department of Physiology, University of Rochester, Rochester, New York (1)*

B. Balke, *Department of Physical Education, University of Wisconsin, Madison, Wisconsin (195)*

H. S. Belding, *Department of Occupational Health, Graduate School of Public Health, University of Pittsburgh, Pittsburgh, Pennsylvania (9)*

W. G. Bradley, *Department of Biological Sciences, University of Nevada, Las Vegas, Nevada (127)*

Robert W. Bullard,* *Department of Anatomy and Physiology, Indiana University, Bloomington, Indiana*

L. D. Carlson, *Department of Human Physiology, School of Medicine, University of California, Davis, California (65)*

C. F. Consolazio, *U.S. Army Medical Research and Nutrition Laboratory, Fitzsimons General Hospital, Denver, Colorado (227)*

F. N. Craig, *Medical Research Laboratory, Edgewood Arsenal, Maryland (53)*

W. H. Forbes, *Harvard School of Public Health, Boston, Massachusetts (191)*

A. Pharo Gagge, *John B. Pierce Foundation Laboratory, New Haven, Connecticut (23)*

S. M. Horvath, *Institute of Environmental Stress, University of California, Santa Barbara, California (183)*

H. L. Johnson, *U. S. Army Research and Nutrition Laboratory, Fitzsimons General Hospital, Denver, Colorado (227)*

R. E. Johnson, *University of Illinois, Urbana, Illinois (99)*

H. J. Krzywick, *U. S. Medical Research and Nutrition Laboratory, Fitzsimons General Hospital, Denver, Colorado (227)*

D. H. K. Lee, *National Institutes of Environmental Health Science, Department of Health, Education, and Welfare, Research Triangle Park, North Carolina (109)*

*Deceased.

U. C. Luft, *Department of Physiology, Lovelace Foundation, Albuquerque, New Mexico (143)*

R. A. McFarland, *Harvard School of Public Health, Boston, Massachusetts (157)*

S. Robinson, *Department of Anatomy and Physiology, Indiana University, Bloomington, Indiana (77)*

Mohamed K. Yousef, *Department of Biological Sciences, University of Nevada, Las Vegas, Nevada (127)*

Preface and Dedication

This book is published as a tribute to Dr. David Bruce Dill, Research Professor and Director of the Laboratory of Environmental Patho-Physiology, Desert Research Institute, University of Nevada System, Boulder City, Nevada in celebration of his eightieth birthday. Many of his students or associates during the last five decades of his life – at the Fatigue Laboratory, Harvard University, Army Chemical Center, Indiana University, and University of Nevada – gathered for a symposium held at the University of Nevada (UNLV) on April 19–20, 1971 to pay him tribute. All were very pleased not only to see him enjoying life in good health but to note his highly active research endeavors carried out six days a week in his new laboratory at the Desert Research Institute, University of Nevada System.

It is our fortune to be reminded that Dr. Dill, an outstanding scholar, has a brilliant record of achievements in the fields of environmental physiology, exercise physiology, and comparative physiology. His contributions have appeared in many scientific journals where one can find over 250 manuscripts. In collaboration with A. V. Bock, Dill wrote the third edition of the "Physiology of Muscular Exercise," published in 1931. This book includes 46 diagrams and 418 references and is considered one of the fundamental foundations of the field of exercise physiology. It was followed by a famous contribution on the physiological responses of living organisms to stress entitled "Life, Heat, and Altitude" published by Harvard University Press in 1938. Recently, with the help of E. F. Adolph and C. G. Wilber, Dill edited the "Handbook of Physiology," Section 4, entitled "Adaptation to the Environment." This book represents the first authoritative reference on the subject and includes 65 chapters written by 66 eminent scientists from many different countries.

The impact of Bruce Dill on the state of physiology is recognized by many colleagues. He has been in the forefront not only as an investigator and teacher but also as an organizer and leader of several expeditions to places such as the Andes, the Canal Zone, and the Colorado deserts. In recognition of his efforts, UNLV and the Desert Research Institute (DRI), University of Nevada System, organized this symposium as a special honor for him. In order to do this, some of his former associates were invited to review the current concepts of physiological adaptations to desert and mountain areas of research to which Dill devoted a great deal of effort and research experience.

We wish to acknowledge the help and support of Dr. Donald H. Baepler, Vice

President for Academic Affairs, UNLV, and of Drs, John M. Ward, President of DRI, Robert B. Smith, Dean, College of Sciences and Mathematics, UNLV, and W. Glen Bradley, Professor of Biology, UNLV, in the organization of the symposium. Dr. Baepler's encouragement and support throughout the preparation of this book are appreciated. The financial support given by the University of Nevada's Board of Regents made this symposium possible.

Mohamed K. Yousef
Steven M. Horvath

David Bruce Dill

Robert W. Bullard

(1929–1971)

Robert Winslow Bullard, distinguished physiologist and associate editor of this volume, died on Mt. McKinley June 24, 1971. He was making physiological observations on the members of the National Outdoor Leadership School of Lander, Wyoming. A group of four, including Bob, had climbed to 9300 feet and, while roped and wearing snowshoes, tramped out an air-drop area. Bob then took off his snowshoes, unfastened the rope, and stepped back on snow that covered a crevasse. A snowbridge collapsed and he fell 130 feet. Postmortem investigation indicated that death was virtually instantaneous.

Born in Waltham, Massachusetts, June 13, 1929, Bob exemplified many traditional characteristics of the New England Yankees. His keen sense of humor, integrity of character, and brilliance in teaching and research were evident to those of us who came to know him in his years at Indiana. After his graduation from Springfield College in Massachusetts in 1951, he received his M.A. degree at the University of Massachusetts in 1954, and his Ph.D. with Edward Adolph at Rochester in 1956. The rest of his career was spent at Indiana University, first at Bloomington and then four years at the Medical School in Indianapolis. In 1962 he returned to Bloomington where from 1964 to 1969 he headed the Department of Anatomy and Physiology. His sabbatical year of 1969–1970 was spent with Hardy in New Haven. During that year he was a Special Research Fellow of the National Institutes of Health. In 1970 he returned to his professorship at Bloomington.

He became Dill's colleague in 1962 when he agreed to write a chapter on Annelids and Molluscs for the volume on "Adaptation to the Environment" that Adolph and Wilber helped Dill edit. That chapter demonstrated his intellectual capacity. In the same year Dill and he were at the Barcroft Laboratory, White Mt. Research Station. While working on different projects they shared the enjoyment of dinner table discussions.

When the National Science Foundation underwrote the establishment of the Laboratory of Environmental Patho-Physiology in 1966 Bob agreed to be a consulting physiologist. In 1968, he and Marlene and their three daughters, Kristen, Carolyn, and Alicia, spent a month in Boulder City, staying at the Hitchin Post Motel. Marlene, a former airline hostess, and the girls proved as adaptable as Bob to the desert environment. Bob's paper on "Responses of the burro to desert heat stress" with Dill and Yousef as joint authors was published in the *Journal of Applied Physiology* in August, 1970. Yousef presented another paper on the two burros (Mabel and Maud) at the Federation meeting in April, 1971. Bob, as chairman, introduced him and suggested an alternate title "Further adventures of Mable and Maud." Bob and Marlene hoped to return with their daughters to Boulder City for the month of August, 1971. Plans for his summer studies were discussed when he came in April to take part in the symposium. His notable contribution to the symposium and his part in the discussion are matters of record. He and Horvath agreed to help Yousef edit the proceedings.

For such a man as Bullard to be cut down at age 42 is a disaster for his family, for his friends, for Indiana University, and for physiology. But we can be thankful that during his few years he left deep and lasting imprints on the family circle, on his many students, and on his colleagues in the fields of environmental physiology, temperature regulation, respiration, and sports medicine.

D. B. Dill
Sid Robinson

I

Some General Concepts of Physiological Adaptations

E. F. ADOLPH

Introduction

The study of physiological adaptations has recurred throughout my research career. In 1920 I spent a summer at a high altitude laboratory in the Alps and saw that mountain climbing became easy after I had lived on Monte Rosa for a couple of weeks. Some years later (1927) I was interested in the possible transition of aquatic animals from freshwater to seawater, and measured the adjustments of frogs to immersion in saline waters (1). In the next decade (1937), I came to Nevada with Bruce Dill and observed that we did the same walk day after day with less and less rise of body temperature and pulse rate (2).

What did these three studies have in common? In each case, exposure to a particular environment modified the body in specific ways. But the specific ways had a uniform pattern in that the modifications increased for a few days and persisted for some days after exposure ceased. Evidently the results of adaptation have long time scales and often enable the body to respond to subsequent exposure in advantageous ways.

At the symposium held when Dill "retired" from Edgewood, Maryland (1961) our theme was physiological adaptation. Upon that and other occasions, I, for one, tried to formulate concepts of physiological adaptations. Now, in 1971, new light can be thrown upon adaptations.

As I see the general problem of physiological adaptations, we need three concepts. First, a precise notion of what these adaptations are; second, a recognition of the similarities among physiological adaptations and tissue hypertrophies, hyperplasias, and regenerations; and third, a comprehensive view of how physiological adaptations and their tissue counterparts are mediated. These I now undertake to explore.

Examples

The following are examples of the modifications induced by environmental or other exposures.

1. Sweat production of men encased in ventilated suits for 2 hr each day increases (Fig. 1). The men are warmed to a uniform oral temperature for the period of exposure; but they sweat more profusely on each occasion (3). Faster sweating enables the men to lose heat more rapidly, provided the sweat evaporates.

2. The number of circulating erythrocytes in men taken to high altitude increases. On successive days erythrocyte production in the bone marrow is increased (4). With more erythrocytes, more oxygen is carried to the tissues.

3. Excretion through the kidneys of various substances such as urea increases when rats are fed a high-protein diet. Not only are the amounts excreted greater, but the clearances also are greater (5). The kidneys attain a new steady output equal to the stepped-up formation of urea.

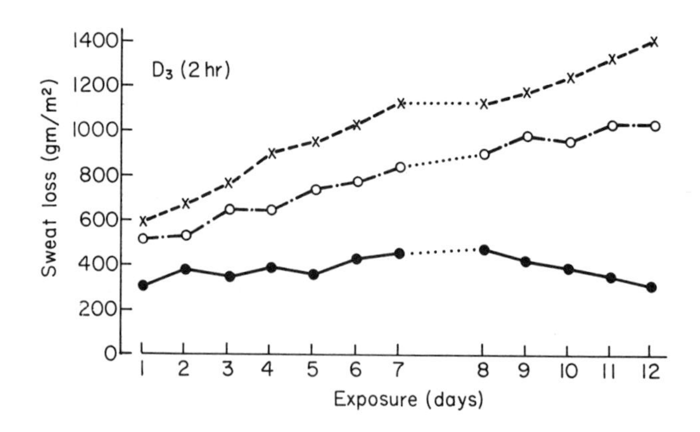

FIG. 1. Mean sweat losses (gm/m² body surface area) for each exposure and treatment group. ●–●, *Human subjects at 37.3°C oral temperature;* o– · –o, *subjects at 37.9°C;* x– – –x, *subjects at 38.5°C. From Fox* et al. *(3).*

Although the word acclimatization might apply to the responses to heat and to altitude, no one would use this term for the buildup of excretory activity. Hence I and others prefer the more general designation physiological adaptation (6).

In each of my three examples I have named one kind of modification. Actually, each environmental shift, which I call an "adaptagent," arouses many measurable modifications, which I call "adaptates." Thus, men arriving at high altitudes undergo dozens of known changes, some of which are represented in Table I (4, 7). To draw attention to their range, I have classified them as structural, chemical, and functional.

But upon analysis one sees that some modifications belong in all classes. From this fact I infer that structural hypertrophies, biochemical inductions, and physiological superfunctions are aspects of one change. From this suggestion I

TABLE I. Adaptates in Man under Hypoxia[a,b]

Changes	*Transient (1st Week)*	*Permanent*
Structural		
Hypertrophies		
Heart mass	→	↑
Erythrocyte volume	↑	↑
Plasma volume	↓	↓
Other		
Hematocrit ratio	↑	↑
Erythrocyte count	↑	↑
Reticulocyte fraction	↑	↑
Vascularity	→	↑
Vital capacity	→	↑
Biochemical		
Hemoglobin in blood	↑	↑
Bilirubin in blood	↑	↑
Glucose in blood	→	↓
Bicarbonate in blood	↓	↓
Functional		
Cardiac output	↑	→
Pulmonary ventilation	↑	↑
Alveolar CO_2	↓	↓
Pulmonary arterial pressure	↑	↑
Pulmonary diffusion capacity	↑	↑

[a] As listed chiefly by Hurtado (4) and Van Liere and Stickney (7).
[b] Arrows indicate increases, no change, or decreases in each adaptate.

find myself looking to hypertrophies for their functional correlatives, and looking to physiological adaptations for their chemical inducers.

This is what I mean by recognition of similarities not only among physiological superfunctions but also among well-known structural hypertrophies and hyperplasias (8).

The concept that superfunction is homologous with hypertrophy doubles our insights into both. The production of superfunction is seen to be as natural as the arousal of hypertrophy.

Kinetics

Each adaptate develops over a time-course of its own (9); a single adaptation creates an overlapping succession of adaptates. Thus, one can follow the acquisition of several modifications in a man kept at simulated altitude for 4 days (Fig. 2). But these modifications are not independent of one another—initial hyperventilation is relieved by subsequent blood changes and a change in erythropoietin output from the kidneys gives rise to an increase in hematocrit

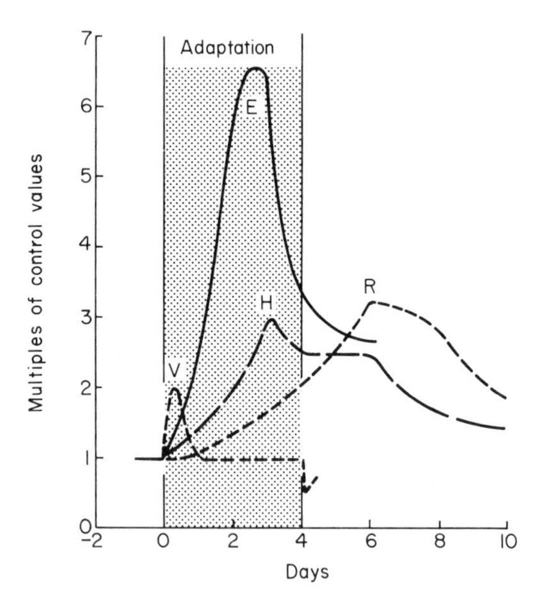

FIG. 2. Semi-diagram of adaptive modifications (adaptates) in a man breathing rarefied air (p_{O_2} 85 mm Hg) during 4 days, followed by 6 additional days of deadaptation. V, lung ventilation; E, serum erythropoietin; H, rate of hemoglobin synthesis; R, fraction of reticulocytes in circulating blood. Data of Siri et al. *(10).*

ratio of circulating blood. Both during and after the exposure to altitude the new adaptates diminish toward the initial state of the tissue and the entire man (10). Each modification is reversible.

One can infer that the capacity to adapt is inherent in every individual. When the appropriate environmental trigger goes off, then that specific adaptation occurs, resulting in the several adaptates that we see in action.

Mediators

Which mediators trigger the adaptations? In several instances, chemical messengers have been identified. They represent one pathway in the response through which superfunction materializes. I here revert to the three examples by which adaptations were originally illustrated.

1. Sweat output is partly controlled by circulating adrenal corticoids. The liberated corticoids may act to diminish the concentrations of sodium and other substances in thermal sweat (11) and probably to increase the flow of sweat whenever a standard heating arouses sweating.
2. The increase in circulating erythrocytes depends in considerable part on the liberation of erythropoietin to stimulate the production of more erythrocytes; erythropoietin is produced as a result of low oxygen tensions in the kidneys (12). Knowledge of a hormone such as erythropoietin calls attention to the specific nature of the means by which adaptation is aroused.
3. Increased flow of urine and decreased sodium reabsorption can be seen as early as 1 hr after one of two rat kidneys has been inactivated (13). Proteins of low molecular weight, such as ribonuclease, when injected, can induce hypertrophy of kidneys and increase the renal clearance. Perhaps the rise of plasma concentration of such proteins that results from the reduced filtration after removal of one kidney arouses the superfunction of the remaining kidney (14).

One may imagine that for each type of superfunction and hypertrophy a particular hormone or group of triggers goes into action, arousing the specific processes by which adaptates become effective.

Functional Demand

Each physiological adaptation appears when the corresponding function is deficient. Thus, an increase in sweating materializes after inadequate body cooling; an increase in erythrocyte production after oxygen deficiency in tissues; an increase in renal clearance after excessive protein metabolism.

The demand for more activity is not a demand for an absolute increment of function (Fig. 3), but represents a signal that output (E, line b of Fig. 3) is deficient relative to the new adaptagent [f(A), line a]. Hence I infer that demand is represented by a ratio (line c) of the actual performance to the new requirement. At first exposure the old performance is small compared to the current f (A); and metabolic deficit accumulates. As time goes on, the increased performance may attain (line c) or not attain (line d) the previous ratio (demand), reaching a steady state little different from the original.

When performance climbs, demand diminishes. The heightened performance itself seems to furnish feedback, thereby reducing the mediators and, hence, certain of the adaptates.

Every tissue has set points of composition and of action. Thus, in a given warm environment, sweat output has an idling speed for one particular body temperature. After adaptation the speed is higher, yet in a new steady state the demand may become the same as when output was idling. Later, deadaptation occurs; the sweat output returns to the original level. Evidently the set point has been preserved throughout the processes of adaptation and deadaptation. In similar fashion the set point appears to be preserved for every function that

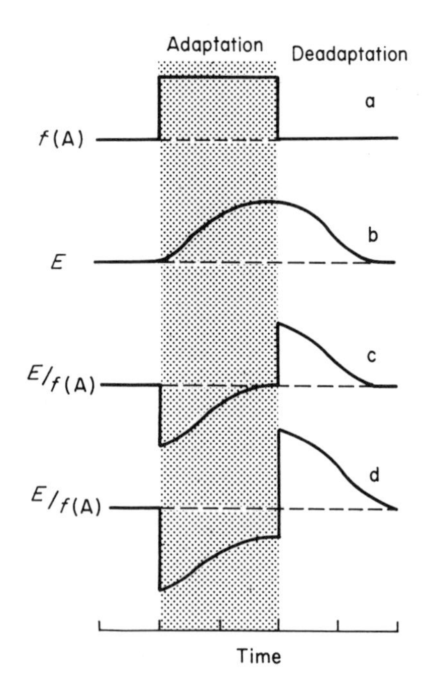

FIG. 3. Diagram suggesting the time-courses of (a) arousal, (b) adaptates, and (c, d) ratio of adaptates to arousal.

undergoes physiological adaptation—the action again finds the constitutional level from which it departed.

Summary

Superfunction is an end product of physiological adaptation. Adaptagents both external to the organism and internal to it arouse adaptive modifications. The modifications tend to restore the ratio between performance and task. Mediated by specific hormones or other messengers, demand often is met and a new steady state is achieved. Such is the pattern of physiological modification that seems to describe what happens when individuals adapt.

References

1. Adolph, E. F. (1927). The processes of adaptation to salt solutions in frogs. *J. Exp. Zool.* **49**, 321.
2. Adolph, E. F. (1938). Heat exchanges of man in the desert. *Amer. J. Physiol.* **123**, 486.
3. Fox, R. H., Goldsmith, R., Kidd, D. J., and Lewis, H. E. (1963). Acclimatization to heat in man by controlled elevation of body temperature. *J. Physiol., (London)* **166**, 530.
4. Hurtado, A. (1964). Animals in high altitudes: Resident man. *In* "Handbook of Physiology" (Amer. Physiol. Soc., D. B. Dill, ed.), Sect. 4, Chapter 54, p. 846. Williams & Wilkins, Baltimore, Maryland.
5. Smith, H. W. (1951). "The Kidney," p. 474. Oxford Univ. Press, London and New York.
6. Dill, D. B., ed. (1964). "Handbook of Physiology," Sect. 4. Williams & Wilkins, Baltimore, Maryland.
7. Van Liere, E. J., and Stickney, J. C. (1963). "Hypoxia," Chapter 10. Univ. of Chicago Press, Chicago.
8. Goss, R. J. (1965). "Adaptive Growth," Chapter 3. Academic Press, New York.
9. Adolph, E. F. (1956). Some general and specific characteristics of physiological adaptations. *Amer. J. Physiol.* **184**, 18.
10. Siri, W. E., Van Dyke, D. C., Winchell, H. S., Pollycore, M., Parker, H. G., and Cleveland, A. S. (1966). Early erythropoietin, blood and physiological responses to severe hypoxia in man. *J. Appl. Physiol.* **21**, 73.
11. Conn, J. W., and Louis, L. H. (1950). Production of endogenous salt-active corticoids as reflected in the concentrations of sodium and chloride of thermal sweat. *J. Clin. Endocrinol.* **10**, 12.
12. Krantz, S. B., and Jacobson, L. O. (1970). "Erythropoietin and the Regulation of Erythropoiesis," Chapter 3. Univ. of Chicago Press, Chicago.
13. Peters, G. (1963). Compensatory adaptation of renal function in the unanesthetized rat. *Amer. J. Physiol.* **205**, 1042.
14. Royce, P. C. (1967). Role of renal uptake of plasma protein in compensatory renal hypertrophy. *Amer. J. Physiol.* **212**, 924.

II

Biophysical Principles of Acclimatization to Heat

H. S. BELDING

Introduction

In northern climates we eagerly await the spring onset of warm weather. Yet when it suddenly arrives we are discomfited. Our customary physical work suddenly is exhausting. Lore has it that our blood is too thick, or we have forgotten how to sweat. Those of us who have been through this know we will not mind the heat so much in a few days. And we call our adjustment acclimatization.

Physiologists have sought and are still seeking to identify mechanisms by which this greater tolerance for environmental heat is achieved. Based on the generalization that man is a homeotherm they have used his ability to maintain normal internal body temperature as a criterion of heat tolerance and of acclimatization. They have focused on the role of the sweating mechanism and on that of the circulatory system in disposing of heat. The effect of acclimatization in facilitating heat transfer and stabilizing body core temperature is the subject of this chapter.

First, I shall briefly review the observed changes in strain that accompany daily exposures to the challenge of activity under heat stress. In examining these adaptations I shall avoid consideration of fundamental changes in nervous and humoral control mechanisms because in the context of acclimatization we know little about them.

DILL'S DEMONSTRATION OF ACCLIMATIZATION

In looking at a complex physiological phenomenon it is interesting to find out what more we know now than was known before we entered the scene. In "Life, Heat and Altitude" (1), we read of relationships between sweating and body temperature. When one Edward Adolph repeated a bout of work in the Nevada desert under approximately the same hot conditions that had exhausted him in wintertime Boston, he found the work easier from day to day. His body temperature rise correlated with his level of fatigue, and after a month this rise following an hour's work was less than half the value of the previous winter, as shown at the left side of Table I, but note that his sweating changed little. On the other hand, when he pushed himself nearly to exhaustion in winter or on certain days in summer, his body temperature rise was always great and independent of his state of acclimatization; his sweat rate increased to the point where it was double that of wintertime. This is shown on the right side of Table I. Dill introduced an observation of C. K. Drinker in discussing these data, namely, "If subjects are exposed to high temperature in the summer their sweating will begin much more rapidly than from the same exposure in winter." Let us leave these three observations for the present and turn to some subsequent work that may provide explanations.

LATER FATIGUE LABORATORY DEMONSTRATIONS OF ACCLIMATIZATION

My own initiation to heat acclimatization, oddly enough, was as a member of a newly assembled cold and high altitude research team at the Fatigue Laboratory (just a few weeks after the bombing of Pearl Harbor). The cold room and altitude chamber were not ready, so Robinson led us into the hot room for a study of the effects of tropical clothing on well-being. What emerged was a serendipitous laboratory demonstration of rapid acclimatization to work in the heat, with data on a very important new parameter, namely skin temperature (2). The data shown in Fig. 1 are final readings available on each day when the subject wore a long-sleeved cotton khaki shirt and full-length trousers.

The plan was to walk at 3.5 mph up a 5.6% grade. Room temperature was set at 40°C and relative humidity at 23%. On the first day we were all forced to stop well short of our 2-hr target, and we were exhausted. Our rectal temperatures at the end were pushing 40°C and our pulse rates were 170-200 beats/min. On subsequent days the strain associated with the same exposures became much less; 20 days later rectal temperatures had fallen to about 38.5°C and pulse rates to 140-160 beats/min. Average skin temperatures had fallen by 1.4°C. The exposures were hardly more difficult than for the same grade of work under cool conditions. To prove a practical advantage of this acclimatization, subjects SR

TABLE I. Body Temperature Rise and Weight Loss of Edward Adolph during One Hour's Work in Heat[a]

		Body weight loss approx. constant		Body temp. rise approx. constant	
		Wt. loss (kg)	*Body temp. rise (°C)*	*Wt. loss (kg)*	*Body temp rise (°C)*
WINTER		0.64	2.20	0.64	2.20
SUMMER					
June	4	0.71	1.79	0.71	1.79
	6	0.67	1.03	–	–
	7	0.69	1.30	–	–
	9	0.67	1.06	–	–
	12	0.61	1.20	–	–
	14	0.74	0.72	–	–
	19	–	–	1.11	1.58
	27	0.71	0.80		
July	1			1.32	2.02
	2			1.32	1.65
	3			1.34	1.68
CHANGE		Little	-1.4	+0.7	Little

[a]Adapted from Dill (1).

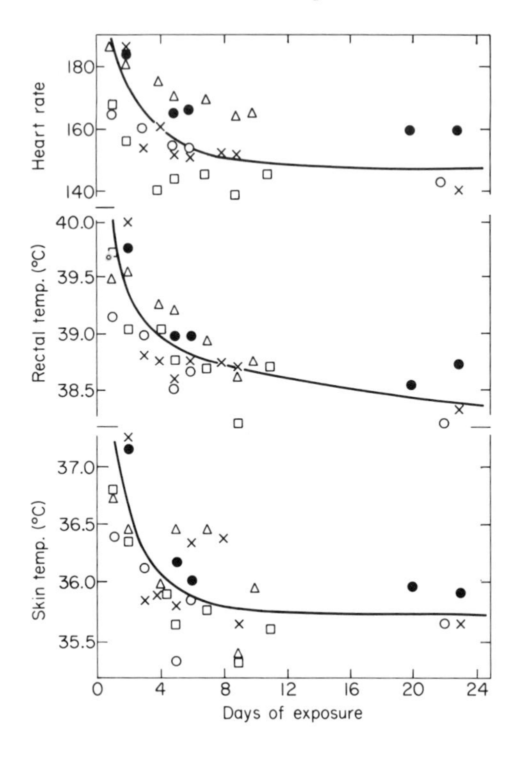

FIG. 1. Heart rate and body temperature at end of treadmill walk on successive days of controlled exposure at 40°C, R.H. 23%. From Robinson et al. *(2)* •, *Subject HB;* ○, *ET;* ×, *SR;* □, *WH;* △, *FC.*

and WH walked for 4½ hr under the conditions that at first had forced termination at 85 min (Fig. 2). In still another demonstration these two subjects completed 3 hr at a work load 35% greater than that during acclimatization.

The importance of skin temperature was not so obvious to us then, because we had only just begun to think in fundamental biophysical terms. Now, owing to the pioneering work of Bazett, Burton, Gagge, Bean, Hatch, and others, we see the implications of our Fatigue Laboratory study more clearly. For example, we can examine the heat transfer situation that we had intuitively selected for the Fatigue Laboratory clothing trials. After acclimatization, with $T_{skin} = {\sim}35.6°C$, the components of heat load are estimated by our best present formulations (3) to have been as follows.

Components of heat load

M	=	520 kcal/hr
W	=	−50
R	=	30
C	=	40
Heat load	=	540 kcal/hr

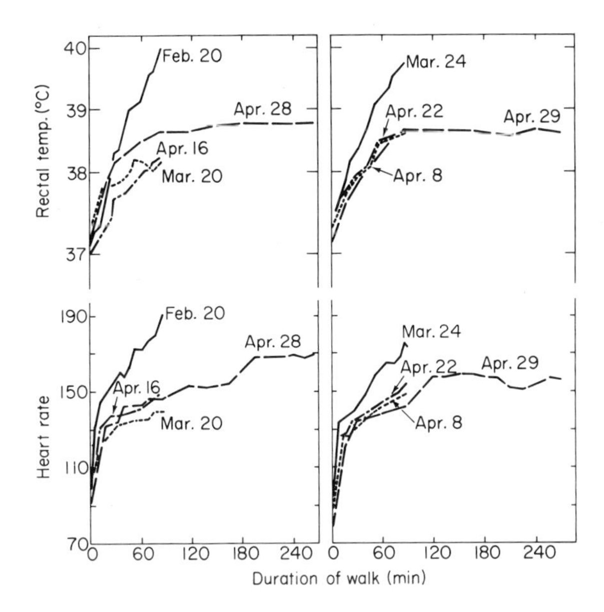

FIG. 2. Time-course of heart rate and rectal temperature of two subjects (left, SR; right, WH) on selected dates of the acclimatization program shown in Fig. 1, plus data from a 4½-hr endurance trial after acclimatization. From Robinson et al. *(2).*

where M is metabolic energy production, W is external work, R is radiation exchange, and C is convective exchange.

To maintain heat balance for the 4½ hr that the two subjects subsequently walked, evaporative cooling had to be at least 540 kcal/hr. This demonstration of acclimatization was enhanced by the fortuitous choice of a combination of work and climate that permitted the achievement of a satisfactory heat balance but only after acclimatization.

The sweat-equivalent of our heat load of 540 kcal/hr is just under 1 liter. These subjects sweated 1.4-1.5 liters /hr to accomplish the necessary wetting of the skin and clothing, an apparent oversweating, which has nevertheless since been shown to be essential when much of the evaporation occurs from the surface of the clothing rather than from the skin. [Under dry conditions, with low ambient pressure of water vapor, and without clothing, such oversweating is not apparent; the sweat rate is in accord with calculated evaporative need (4).]

Using the classic biophysical approach of Burton and Bazett (5) we have computed a "Conductance Index" (CI) value for our Fatigue Laboratory acclimatization study.

$$CI = \frac{M - W}{T_{\text{core}} - T_{\text{skin}}} = \frac{7.8 \text{ kcal/min}}{38.5 - 35.6} = 2.7$$

This *CI* is in the upper range that has been observed for whole-body heat transfer. On the assumption that the specific heat of blood is about 1.0, and neglecting the small fraction of heat lost via the lungs, this means that the requirement for dermal circulation of blood was in excess of 2.7 liters/min (6).

We now compare the average situation on the first day with that after acclimatization which we have just examined.

Before	ACCLIMATIZATION	After
565	M	520
−50	W	−50
20	R	30
30	C	40
565	Heat load	540
	Sweating (liters/hour)	
0.97	Required	0.93
1.43	Actual	1.48
	Dermal gradient (°C)	
39.5	T_{core}	38.5
37.0	T_{skin}	35.6
2.5	ΔT	2.9
3.4	Dermal blood flow required (liters/min)	2.7

The conspicuous differences, aside from the absolute levels of T_{core} and T_{skin}, lie in a 15% greater dermal gradient and a 20% smaller requirement for dermal blood flow.

Mechanisms of Acclimatization

In recent years we have examined mechanisms by which the advantages of acclimatization are achieved. Looking again at our Fatigue Laboratory study, it seems clear that the rectal temperature had to rise higher on the first day because the sweating mechanism was not responsive at a low enough skin temperature, or dermal circulation was inadequate, or both. In other words, sweat rate was not sufficient until skin temperature had reached 37°C, and the cutaneous blood flow that was available was insufficient to unload the metabolic heat without at least a 2.5°C gradient from core to skin. With T_{skin} at 37.0°C,

T_{core} was forced upward to 39.5°C or 40.0°C; this level is associated with rapidly developing exhaustion.

Let us examine separately the contributions of sweating and increased dermal conductance to acclimatization.

SWEATING

I think most students of acclimatization point to increase in sweating rate as a principal explanation for the benefits accrued from lower core temperature and heart rate (7). Yet others, as in the Fatigue Laboratory study, have observed no substantial change in sweating during the process. We believe that such disparate findings can be reconciled. Those who report a substantial increase in sweating have generally used humid heat as the vehicle of stress; for example, Wyndham's group (8) and Ladell (9). We note that with a few exceptions those who have reported little change in sweating with acclimatization (i.e., 5 to 15%) have generally used more arid conditions of exposure (10) (cf. control group of reference 11). Even the small increase may be accountable to the lower T_{skin}, which raises heat gain by $R + C$ (12). Another way of stating this is that sweating was adequate to establish sufficient evaporative cooling from the first day on, but the drive to accomplish this rate was not sufficient until T_{skin} and T_{core} had reached excessive levels.

An hypothesis that accounts for both types of response is simply that use makes the sweating mechanism more responsive; it lowers the skin temperature at which sweating is first elicited and at which any specified amount of sweat is secreted. We shall call this process a *conditioning,* and it does represent an *adaptation,* but I prefer to call it a *quickening* (enlivening) of responsiveness. We cite four demonstrations of this quickening.

The first is our original Fatigue Laboratory study (2). The sweating of the five subjects increased hardly at all, from 1.43 liters/hr before to 1.48 liters/hr after acclimatization. Yet after 3 weeks, this same responsiveness was achieved at a skin temperature of about 35.6°C, instead of the 37.0°C at the end of the first exposure; and it was achieved at a rectal temperature of about 38.5°C, rather than at the 39.5°-40.0°C that was reached during the first exposure.

A second demonstration is based on Hatch's analysis of data from the comparison of unacclimatized subjects at Oxford, England, and naturally acclimatized subjects of Singapore (13). The latter had a threshold of sweating response at a skin temperature roughly 2°C lower than those at Oxford. And at any skin temperature higher than threshold they sweated at least 0.5 liters/hr more than the Oxford men.

A third demonstration derives from our study of responses of five previously unacclimatized men who cycled for 2 hr daily while immersed to the neck in a water bath at 37.1°C. Skin temperature was clamped by the bath at 37.1°C. After 14 days the sweating rate (from weight loss corrected for water intake) at the same T_{skin} had nearly doubled (Fig. 3a).

The fourth demonstration of quickening is attributable to Fox *et al.* (14). After daily local heating of one arm, its sweat production in a first heat exposure of the whole body was much greater than that of the opposite, previously unheated arm. Likewise, when one arm was maintained cool during 15 days of whole-body heat acclimatization, that arm sweated less than the other on subsequent exposure. These observations were considered consistent with the hypothesis that "the increase in sweating which occurs during heat acclimatization is primarily a local training response of the sweat glands and not the result of an increased intensity of stimulation by the C.N.S." This statement may provide a lead for any who would like to achieve rapid acclimatization without submitting subjects to prolonged elevation of body core temperature, i.e., try quickening the sweating by intermittent strong heating of the skin.

Quickening helps us to reconcile and explain the changes brought about by acclimatization in both dry and wet heat. From the first day in dry heat the sweat rate is sufficient or very nearly sufficient to meet the evaporative need, but it takes an excessively high skin and/or core temperature to drive the mechanism. Adequate evaporation is no problem. After acclimatization, about the same amount of sweat suffices, but this is produced at a 1-2°C lower skin temperature because of the quickening phenomenon. The lower skin temperature in turn means that core temperature is not pushed to such a high level, and the increased core-to-skin gradient that is available even reduces the requirement for dermal circulation as shown in our model based on the Fatigue Laboratory study.

On the first day in wet heat it seems likely that a tolerable physiological tax in terms of elevated skin and/or core temperature does not produce enough sweat to wet the skin fully. (Because of unequal distribution, 100% wetting of all the skin apparently requires up to 50% more sweat than can be evaporated.) Subsequently, as a result of quickening, the effective wetted area is increased; this in turn permits equilibrium at a somewhat lower skin temperature. However, T_{skin} cannot fall to the extent observed in dry heat because the water vapor pressure generated at the skin, which is dependent on T_{skin}, would otherwise fall below the critical level for achieving adequate evaporative cooling. The sweat rate in wet heat is higher after acclimatization because of both (1) quickening and (2) the necessarily high T_{skin}.

DERMAL CONDUCTANCE

Changes in circulating blood volume and in allocation of blood among vital organs probably improve venous return and lessen circulatory strain for the

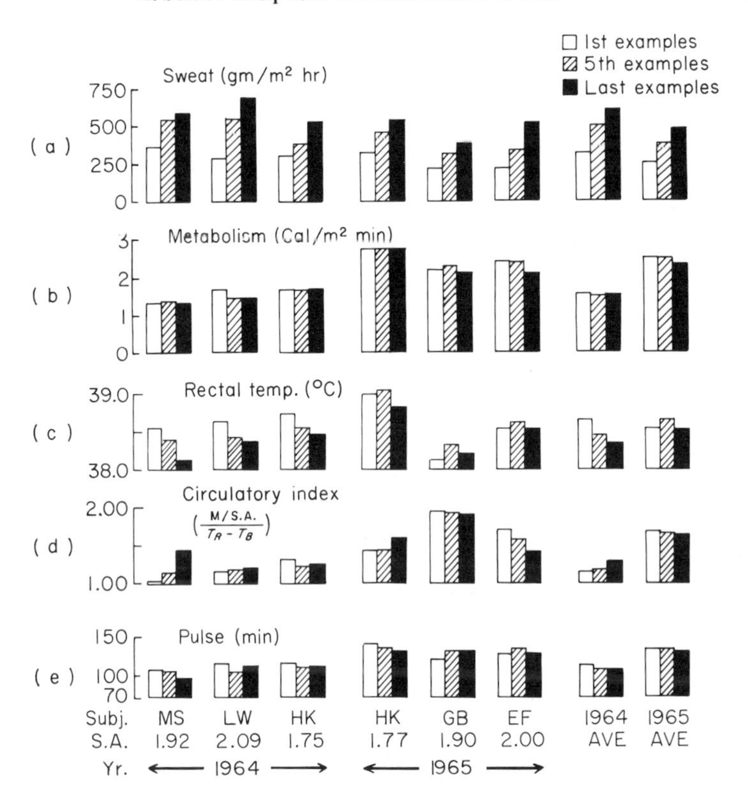

FIG. 3. Responses of five unacclimatized subjects to 14 daily 2-hr early spring exposures in a water bath at 37.1°C and pedaling. (a) Rate of sweating nearly doubled in 15 days; (b) metabolic rate did not change, but the work load was higher in 1965 than in 1964; (c) rectal temperature changed for some individuals, but not significantly for the group; (d) circulatory index was higher with the higher work rate but did not change with acclimatization; (e) pulse rate was higher with the higher work rate but did not change with acclimatization.

acclimatized. These adjustments may well account for some lowering of heart rate and for the lessened likelihood of heat exhaustion after heat acclimatization. However, it is also reasonable to suppose that improved thermal conductance, achieved through greater or better distributed blood flow through the skin, also accounts in part for the improved well-being of acclimatization (7). Here we confine ourselves to consideration of evidence on changes in this regard. In our view the evidence for improvement in dermal conductance as a feature of acclimatization is scant, though Fox *et al.* (15) were able to enhance forearm blood flow (largely dermal) by a program of whole body acclimatization. We have made three different efforts to demonstrate such a change in *CI*; each was futile.

In the first instance the water bath immersion previously cited was utilized. In performing standardized work in the water bath at 37.1°C the metabolic heat production could be balanced only after rise in body core temperature had occurred sufficient to establish an adequate gradient from core to skin. Change in sweat rate would avail for evaporative cooling in this situation only to the extent that it occurred from the head, and this avenue was restricted by the high vapor pressure above the surface of the water. The hypothesis was that with acclimatization the elevation of core temperature would become less. Thus $M/(T_{core} - T_{skin})$, i.e., *CI*, would become greater, indicating more effective heat transfer. The first three subjects who were exposed in 1964 had slightly lower average T_{core} after 15 exposures, but this finding was not supported when three subjects were studied in 1965 (Fig. 3c). Their heart rates did not show evidence of decreasing strain either (Fig. 3e). Although we failed to find consistent support for our hypothesis, it was true that each individual subject had a characteristic *CI* value and that this varied considerably among individuals (Fig. 3d). A single exposure by this procedure might provide a measure of circulatory fitness for heat, and we suggested exploration of this possibility in connection with the International Biological Program (16).

There was another finding in this study. When subjects who had undergone the series of exposures in the water bath then walked in dry heat in our exposure chamber their responses were those of acclimatized men. According to the usual indices little subsequent improvement was obtainable (Fig. 4). A heated swimming pool might be the medium of choice for preacclimatization of large groups of men.

Our second attempt to demonstrate change in *CI* involved only one unacclimatized subject; he worked in a very humid air environment (37°C DB/34.5°C WB) on ten successive days. Here also there was no evidence of change in *CI*. (Nor did we see much lowering of heart rate or rectal temperature.) As in the water bath study, the substantial change was in sweating rate; this rose from about 1.2 to 2.2 liters/hr over the 10-day period.

Our third attempt at demonstrating change in *CI* with acclimatization was made on hands and forearms, using a water calorimeter on the same subjects, in winter and summer. Here the temperature gradient was reversed, as described in reference 6. Heat uptake of the limb segment from a thermostatted bath that was warmer than core temperature was the measurement of interest. There was no seasonal difference in this heat uptake (Fig. 5), even though the subjects had been actively acclimatized prior to the summer tests.

Each of these three experimental protocols can be faulted. But as it stands now, we are forced to attribute most of the benefit of acclimatization to adaptation of the sweating mechanism rather than to improved capacity for transmitting metabolic heat through the skin.

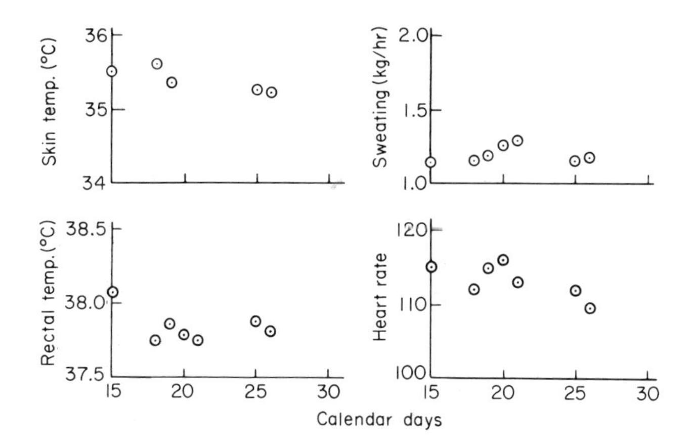

FIG. 4. Responses at end of work periods in dry heat on days immediately following 14 days of work in the bath at 37.1°C. Values of the usual indices are close to plateau levels, indicating that the men had achieved nearly complete acclimatization from their previous exposures in wet heat. Average data of three subjects walking 5.6 km/hr; 50°C dry bulb, 23.5°C wet bulb.

Summary

In conclusion, the weight of evidence points to quickening of the sweating mechanism as the major factor in acclimatization to heat. In dry heat this permits lower skin and core temperatures with small change in rate of sweating. In wet heat, the quickening results in a markedly increased sweat production, which effectively assures a larger wetted area of skin and thereby greater evaporative cooling.

Evidence that an increase in dermal conductance plays any important part in acclimatization to heat is scant. The decrease in circulatory strain in dry heat after acclimatization can be attributed in large part to the lower skin temperature. In wet heat it seems to result primarily from the better balancing of the circulation against competing demands of muscle, vital organs, and skin.

Postscript

We left hanging our interpretation of Dill's findings on Adolph which were mentioned at the beginning of this chapter (Table I). In the exposures involving

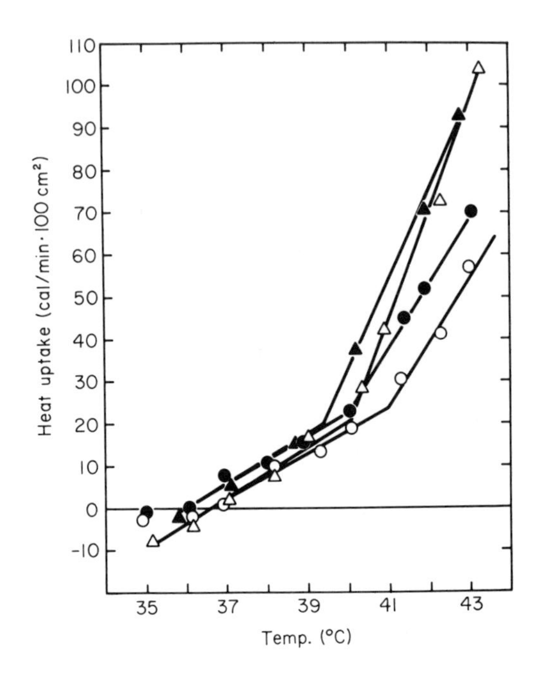

FIG. 5. Average heat uptake of hand and forearm of three subjects at various calorimeter temperatures. Hand: △, *summer;* ▲, *winter. Forearm:* ○, *summer;* ●, *winter. The values for summer and winter are not different.*

equivalent stress from day to day during acclimatization, body temperature could fall from day to day because of quickening of the sweating mechanism. It did not require appreciably more sweat production to achieve this. Drinker's observation, that sweating begins more rapidly in summer, is further evidence of this quickening.

In the series of exposures that produced exhaustion each day, the work or climate was more stressful from day to day. Maximum capacity of the sweating mechanism was being tested each day, and this capacity was increasing markedly with acclimatization.

References

1. Dill, D. B. (1938). "Life, Heat and Altitude." 211 pp. Harvard Univ. Press, Cambridge, Massachusetts.
2. Robinson, S., Turrell, E. S., Belding, H. S., and Horvath, S. M. (1943). Rapid acclimatization to work in hot climates. *Amer. J. Physiol.* **140**, 168.
3. Belding, H. S. (1967). Resistance to heat in man and other homeothermic animals. *In* "Thermobiology" (A. H. Rose, ed.), p 479. Academic Press, New York.
4. Belding, H. S., Hertig, B. A., and Kraning, K. K. (1966). Comparison of man's responses to pulsed and unpulsed environmental heat and exercise. *J. Appl. Physiol.* **21**, 138.

5. Burton, A. C., and Bazett, H. C. (1936). A study of the average temperature of the tissues, of the exchange of heat and vasomotor responses in man by means of a bath calorimeter. *Amer. J. Physiol.* **117**, 36.
6. Kamon, E., and Belding, H. S. (1968). Heat uptake and dermal conductance in forearm and hand when heated. *J. Appl. Physiol.* **24**, 277-281.
7. Bass, D. E. (1963). Thermoregulatory and circulatory adjustments during acclimatization to heat in man. *Temp.: Its Meas. Contr. Sci. Ind., Proc. Symp., 4th, 1961* Vol. 3, Part 3, p. 299. Reinhold Publishers, New York.
8. Peter, J., and Wyndham, C. H. (1966). Activity of the human eccrine sweat gland during exercise in a hot humid environment before and after acclimatization. *J. Physiol. (London)* **187**, 583.
9. Ladell, W. S. S. (1947). Effects on man of high temperatures. *Brit. Med. Bull.* **5**, 5.
10. Bean, W. B., and Eichna, L. W. (1943). Performance in relation to environmental temperature. *Fed. Proc., Fed. Amer. Soc. Exp. Biol.* **2**, 144.
11. Buskirk, E. R., Iampietro, P. F., and Bass, D. E. (1958). Work performance after dehydration: Effects of physical conditioning and heat acclimatization. *J. Appl. Physiol.* **12**, 189.
12. Lind, A. R., and Bass, D. E. (1963). Optimal exposure time for development of acclimatization to heat. *Fed. Proc., Fed. Amer. Soc. Exp. Biol.* **22**, Part 1, 704.
13. Belding, H. S., and Hatch, T. F. (1963). Relation of skin temperature to acclimation and tolerance to heat. *Fed. Proc., Fed. Amer. Soc. Exp. Biol.* **22**, 881.
14. Fox, R. H., Goldsmith, R., Hampton, I. F. G., and Lewis, H. F. (1964). The nature of the increase in sweating capacity produced by heat acclimatization. *J. Physiol. (London)* **171**, 368.
15. Fox, R. H., Goldsmith, R., Kidd, D. J., and Lewis, H. E. (1963). Blood flow and other thermoregulatory changes with acclimatization to heat. *J. Physiol. (London)* **166**, 548.
16. Belding, H. S., Hertig, B. A., Kraning, K. K., Roets, P. P. and Nagata, H. (1966). Use of a water bath for study of human heat tolerance. *In* "Human Adaptability and Its Methodology" (H. Yoshimura and J. S. Weiner, eds.), p. 115. Japan Society for the Promotion of Sciences, Tokyo.

III

Partitional Calorimetry in the Desert

A. PHARO GAGGE

Introduction

A dry desert environment as we have in Las Vegas is usually characterized by widely varying diurnal ambient temperatures, a relatively constant dew point temperature, air movements that tend to increase with ambient temperature, daytime solar heat loads with magnitudes up to three times man's sedentary metabolic heat, and nighttime negative radiant cooling to the clear sky. Except for an increased emphasis on these environmental factors, the partitional calorimetry involved for man is essentially the same we would use for any other living environment.

Since the early beginning of partitional calorimetry (1, 2), there have been a long stream of ideas, simplifying concepts, and detailed physical analyses of the heat exchange between man and his environment. The present mathematical but bioengineering-oriented description of man's environmental heat exchange includes concepts in recent fundamental papers by Fanger (3,4), Rapp (5), and Nishi (6-8), all of whom have updated many conventional practices in partitional calorimetry. No study of environmental heat exchange can avoid the use of the earlier fundamental concepts proposed by Hardy (9) and Burton and Edholm (10). Every contributor to this field has had his own personal way of looking at the environment. The study of the human body is an inexact science and the use of compromising assumptions will continue to be a necessity for a long time to come. The following presentation is a summation of the quantitative data currently available and of use in dynamic partitional calorimetry. Each equation is written as a *definition* of some physical or physiological concept. The ultimate objective of partitional calorimetry is a definition of the state of the body's thermal equilibrium in terms of the key physiological and physical factors involved. Details on laboratory methods of measurements are omitted, because they are well known; however, newer methods of direct measurement will be indicated when pertinent.

The Independent Variables

FOR MAN

The principal independent physiological variable for a man is the metabolic rate M produced by his chosen activity. M is expressed in watts and, preferably, per unit body surface area measured by the Dubois formula (11) in square meters.

$$A_D = 0.203\,(w)^{0.425}\,(h)^{0.725} \tag{1}$$

where w is body mass in kilograms and h is height in meters. Associated with M there is a second independent variable, the work accomplished (W) during his activity.

FOR THE ENVIRONMENT

There are seven independent variables.

Ambient Air Temperature

The ambient air temperature is expressed as T_a in degrees Centigrade.

Dew Point Temperature

The dew point temperature is expressed as T_{dew} in degrees Centigrade. Alternate measures of humidity, dependent on T_a, are relative humidity (ϕ_a) expressed in this case as a fraction, and wet bulb temperature (T_{wet}). The saturated vapor pressure P_{dew} at dew point temperatures T_{dew} is the ambient vapor pressure in millimeters of mercury. Of the five variables associated with humidity, namely T_{dew}, ϕ_a, T_{wet}, T_a, and P_{dew}, if any two are measurable the other four may be found by using a psychometric chart (12) or may be calculated (13) by solving the following equations by iteration

$$\phi_a = [P_{wet} \quad 0.00066 \cdot B \cdot (T_a - T_{wet})(1 + 0.00115 T_{wet})] / P_a \tag{2}$$

and

$$P_{dcw} = \phi_a P_a \tag{3}$$

where P_a is saturated vapor pressure at T_a, B is the barometric pressure in millimeters of mercury. Tables relating the saturated vapor pressure (P) (mm/Hg) at temperature °C may be found in reference 13. An accurate analytic equation relating saturated vapor pressure and temperature and useful for on-line computer analysis of observed data has been recently developed by Spofford (14).

Air Movement

Expressed as v in meters/second, the air movement affecting the subject is of two types—(a) natural, and (b) forced. The latter may be caused by the ambient air movement itself as well as by the bodily motions of the subject during his activity. Air movement affects both the convective heat transfer and the mass transfer by evaporation.

Thermal Radiation

Either a temperature or an energy mode may be used for describing the variables associated with radiation exchange.

Mean Radiant Temperature. $\bar{T}_r$ is defined (15) as the temperature of an imaginary "black" enclosure in which man will exchange the same heat by radiation as in his actual nonuniform environment.

Effective Radiant Field (ERF). The ERF is defined (16) as the radiant energy (W/m^2) exchanged by man with the above imaginary "black" enclosure at $\bar{T}_r$ if his body surface temperature were considered the same as the ambient air temperature. Only those surfaces differing in temperature from the ambient temperature contribute to the (ERF) affecting man.

Clothing Insulation

For clothing insulation, I_{clo} is the unit to be used and is defined as $0.155°C \cdot m^2/W$.

Barometric Pressure

This factor (B) in millimeters of mercury affects heat transfer by both the processes of convection and evaporation.

Time of Exposure

The time of exposure (t) to above six variables is expressed in hours.

The Dependent Variables

The physiological variables associated with man's heat exchange with his environment are classed here as dependent variables. They will be limited here to the following four.

Mean Skin Temperature

The measurement $\bar{T}_{sk}$ is an average of at least eight local values of T_{sk}, weighted by the fraction of the total body surface represented by segmental T_{sk} concerned. A useful weighing scale (17) is head (7%), chest (17.5%), back (17.5%), upper arms (7%), forearms (7%), hands (5%), thighs (19%), and legs (20%).

Central Core Temperature

The direct measurements that may serve as indices of core temperature T_{cr} are rectal T_{re}, esophageal T_{es}, tympanic T_{ty}, and muscle T_{m} (e.g., quadriceps).

Total Evaporative Heat Loss

The total loss, E, in watts per square meter, is caused by (1) respiration from the lung, (2) diffusion through the skin surface layer, and (3) regulatory sweating on the skin surface.

Skin Blood Flow (SKBF)

Expressed in liters per hour per square meter of skin surface, the SKBF is a hypothetical measure of the heat transfer caused by blood, circulating from the central core of the body to its skin surface, and is derived from direct measurements of $\bar{T}_{sk}$, T_{cr}, and the total heat flow from the skin surface to the environment.

How the above independent variables and four dependent variables are interrelated during the passive state existing at any time t is the initial objective of this chapter. The ultimate purpose of partitional calorimetry is to measure directly all transient temperature and energy changes and to understand the relationships between the four dependent physiological variables themselves that best describe human body temperature regulation.

The Passive State

The heat balance equation describing the passive state takes the classic form

$$S = M - (E \pm W) \pm R \pm C, \quad \mathrm{W/m^2} \tag{4}$$

where S is rate of heat storage; M, metabolic rate; W, rate of work accomplished; E, rate of total evaporative heat loss; R, radiation exchange; and C, convective heat exchange. Positive (+) values for each term raise body temperature; negative (–) lower.

For the present analysis we will consider the outer skin surface to be the boundary separating man's body from his environment, and from which the exchange of heat with environment is concerned. We will define the term *net metabolic rate*, M_{net}, to describe the net heat flow from the internal body to its skin surface. We will assume all evaporative heat loss E_{sk} occurs on the skin surface. Finally, we will assume the dry heat exchange with the environment from the clothing surface to be the same as the dry heat flow from the skin surface through the clothing itself. Our revised heat-balance equation, defining the rate of heat storage (S) becomes

$$S = M_{net} \pm R \pm C - E_{sk} \quad \mathrm{W/m^2} \tag{5}$$

The usefulness of partitional calorimetry lies in our ability to measure accurately the four terms on the right in Eq. (5). Each term will now be discussed individually.

Net Metabolic Rate (M_{net})

There are three avenues through which the metabolic energy may leave the body without passing through the skin surface.

1. Work performed, W, which can be measured on an ergometer or treadmill by conventional methods.

2. The respired vapor heat loss E_{res}, which has been shown by Fanger (3, 4) to be given by

$$E_{res} = 0.0023\,M\,(44 - P_{dew}) \tag{6}$$

This relation, useful only up to metabolic rates approximately 6 times that of resting, is a bioengineering simplification of the original formula developed by McCutchan and Taylor (18).

3. The respired convective heat loss C_{res} (4), is given as

$$C_{res} = 0.0012\,M\,(34 - T_a) \tag{7}$$

By combining the three factors, the net metabolic heat M_{net} is defined by

$$M_{net} = M[1 - 0.0023(44 - P_{dew}) + 0.0012\,(34 - T_a)] - W \tag{8}$$

M is determined from the aerobic oxygen consumption.

In general positive work ($+W$) while on a bicycle ergometer, tends to reduce M_{net}; negative work ($-W$), while walking down steps, tends to increase M_{net}.

Clothing in Heat Exchange

We will follow the original suggestion by Burton (10) and describe the role of clothing in heat exchange—both sensible and insensible—by two dimensionless factors, which modify the flow of dry heat and the mass transfer of water vapor from the skin surface to the environment.

In its simplest form Burton's "thermal efficiency" factor F_{cl} is described in the dry heat exchange terms $(R + C)$ or DRY, when $\overline{T}_r = T_a$, by

$$\text{DRY} = h(\overline{T}_{cl} - T_a) \tag{9}$$

or

$$= h_{cl}(\overline{T}_{sk} - \overline{T}_{cl}) \tag{10}$$

and

$$= h(\overline{T}_{sk} - T_a)\,[h_{cl}/(h + h_{cl})] \tag{11}$$

where h is the combined heat-transfer coefficient from the clothing surface at mean temperature $\overline{T}_{cl}$ to the ambient air T_a, and h_{cl} is the conductance of the clothing from $\overline{T}_{sk}$ to $\overline{T}_{cl}$. The units for both are $W/(m^2 \cdot {}^\circ C)$. Thus

$$F_{cl} = h_{cl}/(h + h_{cl}) \quad \text{N.D.} \tag{12}$$

or

$$= I_a/(I_a + I_{cl}) \tag{13}$$

where $I_a = 1/h$ and $I_{cl} = 1/h_{cl}$. In terms of I_{clo} the clothing insulation in clo units is given by

$$F_{cl} = 1/(1 + 0.155 \cdot h \cdot I_{clo}) \quad \text{N.D.} \tag{14}$$

The corresponding factor for vapor transfer through the clothing is called "permeation efficiency" (F_{pcl}). By recognizing the parallel relationship between temperature (°C) and vapor pressure (mm Hg) gradients from skin surface to ambient air, then

$$F_{pcl} = h_{ecl}/(h_e + h_{ecl}) \tag{15}$$

where h_e is the evaporative heat transfer from the clothing surface to the ambient air (analogous to h) and h_{ecl} is the vapor heat transfer through the clothing (analogous to h_{cl} or $1/I_{clo}$). In terms analogous to Eq. (14)

$$F_{pcl} = 1/(1 + 0.143 \cdot h_c \cdot I_{clo}) \quad \text{N.D.} \tag{16}$$

where h_c is the convective heat transfer coefficient. Equation (16) has been derived (6, 7) on the basis of mass transfer theory and has been validated experimentally for *normal porous fabrics* by naphthalene simulation. Use of this simplified concept for F_{pcl} assumes that all evaporation occurs on the skin surface, and that the vapor pressure gradient through the normal porous clothing worn is continuous. Equation (16) does not apply to vapor impermeable clothing; as may be seen from Eq. (15), such clothing causes h_{ecl} to approach zero and hence F_{pcl} approaches zero.

Radiation Exchange

The radiation component R of the dry heat exchange in desert calorimetry is best described by the energy mode which in its simplest form may be written (16) as

$$R = (\text{ERF})F_{cl} - h_r(\overline{T}_{sk} - T_a)F_{cl} \quad \text{W/m}^2 \tag{17}$$

where

$$h_r = 4\sigma[(\overline{T}_{cl} + T_a)/2 + 273]^3 \times (A_r/A_D) \times (1 + 0.15 \cdot I_{clo}) \tag{18}$$

and

$$\overline{\overline{T}}_{cl} = T_a + F_{cl}(\overline{T}_{sk} - T_a) \tag{18a}$$

σ = the Stefan-Boltzmann constant or 5.67×10^{-8} W/(m^2 ·°K^4), and A_r/A_D = the ratio of the 4π radiating area of the human body surface to its Dubois surface area.

A_r/A_D has recently been determined directly with considerable accuracy by Fanger (4), using optical methods, and is found to vary from 0.70 for the sitting position to 0.725 for the standing position within ± 2%, regardless of height or body weight. He has further shown as indicated in Eq. (18) that the ratio A_r/A_D increases by approximately 15% for each clo unit of insulation worn. In Eq. (17) the last term on the right describes the radiant heat exchange from the skin surface to a uniform environment at ambient temperature T_a.

The Effective Radiant Field is the sum of the energy exchanges with all sources, whose radiating temperatures differ from T_a. Thus,

$$(\text{ERF}) = (\text{ERF})_1 + (\text{ERF})_2 + (\text{ERF})_3 + \ldots + (\text{ERF})_n \tag{19}$$

$(\text{ERF})_1$ may represent the direct solar radiation absorbed; $(\text{ERF})_2$ radiation reflected from the sky, $(\text{ERF})_3$ radiation reflected upward from the ground surface, $(\text{ERF})_4$ long wave radiation emitted by the surrounding warm ground surfaces, $(\text{ERF})_5$ may be alternatively the negative radiation lost to a "cold" night sky. In general, any $(\text{ERF})_n$ is a function of the radiating temperature of the source n in degrees Kelvin, the absorptance α of the exposed body or clothing surface for the emission σK^4 of the source, and an appropriate shape factor F_{m-n} that describes the fraction of man's total body surface irradiated by the source. Any radiant source n contributing to the total ERF can take the generalized form.

$$(\text{ERF})_n = F_{m-n}[\alpha_n \sigma (T_n + 273)^4 - (T_a + 273)^4] \quad \text{W/m}^2 \tag{20}$$

Figure 1 illustrates how the absorptance α for human skin varies with the temperature spectrum of the radiating source (16, 19).

The shape factor $F_{m\text{-}n}$ as here used, refers to the total body area A_D, rather than the radiating area of the body A_r. Only for the simplest geometrical arrangements are values of $F_{m\text{-}n}$ easily evaluated. If the radiating source is an infinite sphere, $F_{m\text{-}n}$ would be 0.70-0.73, as we mentioned above. At the center of an infinite hemisphere (e.g., the night sky) $F_{m\text{-}n} \sim 0.35\text{-}0.37$. For a point source such as the sun at a 60° angle, $F_{m\text{-}n} \sim 0.2\text{-}0.25$, depending on posture. In every case, $(\text{ERF})_n$ is the energy absorbed on the outer surface of the body and is a function of the absorptance of the skin or clothing surface, but is *not* a function of its temperature but T_a instead.

Roller and Goldman (20) have recently published a comprehensive study on the solar load for clothed soldiers working in various geographical climates. By using their sample data as a guide, we can evaluate the case for a person walking

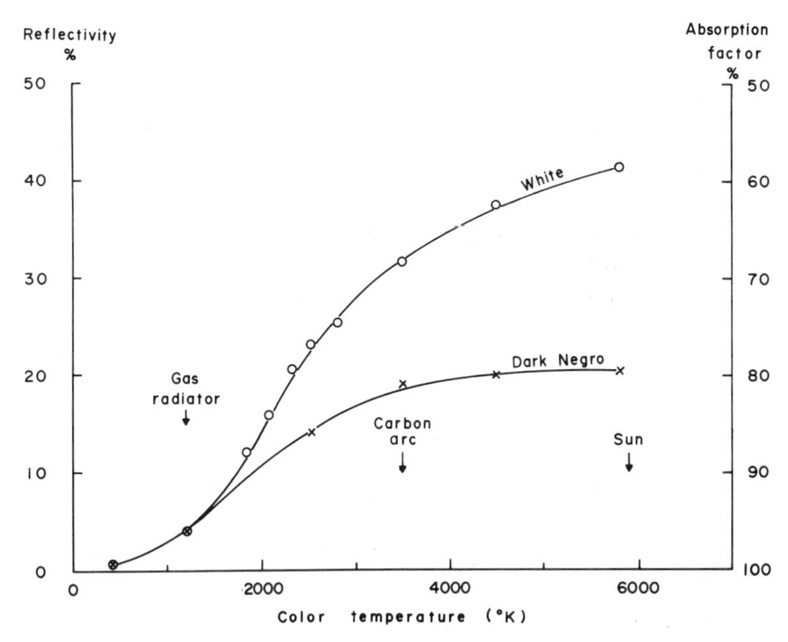

FIG. 1. The variation of absorptance (α) of human skin as a function of the radiating temperature of the source (19). Total surface reflection and absorption of human skin for black body radiators.

in the desert when the sun is at a 60° angle above the horizon. In Eq. (20) the irradiation received from the sun at temperature T_n in degrees Kelvin is covered by the energy term $\sigma(T_n)^4$; the term with $T_a{}^4$ may be ignored. The total solar load on a clear day consists of three parts (1) the direct (~ 800 W/m^2); (2) the diffuse (~ 152 W/m^2); and (3) the albedo or energy emitted and scattered by the terrain (~ 107 W/m^2). The respective angle factors ($F_{m\text{-}n}$) for each case are 0.24; 0.48; and 0.36. If α for either skin or clothing surface worn is 0.7, the total solar load absorbed on the body or clothing surface for (1), (2), and (3) would be 212 W/m^2 for ERF. For $h = 20$ $W/(m^2 \cdot °C)$, $I_{clo} = 0.5$ (i.e., $F_{cl} = 0.39$) the net load on the skin surface would be 85 W/m^2. Without clothing it is clear that the solar load on the bare skin surface could have been the 212 W/m^2.

In practice it would be desirable to measure the total ERF received on the body surface directly by a 4π radiometer. Although many sophisticated instruments for measuring total radiant heat have been made in the past [the Pan-Radiometer (21), the Two-Sphere (22), and R-meter (23)], they are either extinct or one-of-a-laboratory kind. The simplest direct measure of (ERF) for man generally available today is by a *skin-colored* Bedford globe thermometer (6-in. diameter) and by using the following formula

$$(\text{ERF}) = A_r/A_D\,[6.1 + 13.6\sqrt{v}]\,(T_g - T_a) \quad \text{W/m}^2 \tag{21}$$

where v is the ambient air movement in meters per second. The first term in brackets is the value of h_r for a sphere at average temperature 27°C. The second term is Bedford's formula for the convective heat loss from a 6-in. globe, when converted to meters per second and degrees Centigrade. The ratio, A_r/A_D or 0.72, converts what radiation field the sphere "sees" to what man "sees" for a 4π solid angle. The *skin color* allows the globe to absorb radiant heat as man does regardless of the energy spectrum. If the globe were matte black ($\alpha \sim 1$), the energy calculated by Eq. (21) must be multiplied by α (i.e., 0.7) if the sun were the principal source.

Convection

The heat exchange by convection is given by

$$C = h_c(T_a - \bar{T}_{sk})\,F_{cl} \cdot (B/760)^{0.55} \quad \text{W/m}^2 \tag{22}$$

where h_c is the convective heat-transfer coefficient from the outer body surface. The coefficient h_c has proven in the past to be one of the most difficult to measure accurately. The usual calorimetric measurements of h_c require the certainty that either body-heat storage S is zero or can be measured accurately, that careful measurements of the always associated radiation are possible. The value for h_c when obtained by these methods serves primarily as a calibration coefficient for the experimental arrangement concerned.

The physical factors that contribute to h_c are not simple (5). Conditions of free and forced convection may exist simultaneously. The forced air motion surrounding the subject may be laminar or turbulent; both may occur simultaneously and may be caused by the motion of the ambient air as well as by activity on the part of the subject. The exact relationship of the coefficient h_c to the ambient air movement v may be different for turbulent and laminar flow. Finally the question remains as to how the air movement v itself about man is measured accurately and consistently.

Measurements of h_c for various standard activities used in the laboratory are presented in Tables I and II. Those reported by Nishi (8) represent the first *direct* measurements of h_c during exercise, treadmill walking, and free walking without the use of calorimetry. The last equation for h_c, combining the effects of free walking and ambient air movement in the opposite direction, is based on a comparison of the h_c coefficients for free and treadmill walking, the difference being the contribution of ambient air movement. The values presented have an experimental basis up to speeds of 4 mph (or 1.8 m/sec). For greater velocities these equations are only useful as first-order planning estimates.

Table I. Convective Heat-Transfer Coefficients in Normally Ventilated Environments at Sea Level

Condition	h_c W/(m^2 · °C)	*Ref.*	*Remarks*
Seated	2.9 ± 0.9^a	(1, 24)	Partitional calorimetry
Standing	4.5 ± 0.3^a	(24)	Partitional calorimetry
Reclining	4.3 ± 0.9^a	(25)	Longitudinal flow
Seated on stool	3.1^a	(8)	Naphthalene method
Seated on bicycle	3.4^a	(26)	Partitional calorimetry
Pedaling bicycle ergometer	4.8 ± 0.8	(26)	Partitional calorimetry
Pedaling bicycle ergometer (50 rpm)	5.4	(8)	Naphthalene method
Pedaling bicycle ergometer (60 rpm)	6.0	(8)	Naphthalene method

[a]These values involve both free and forced convection and may be considered as constant and representative of ambient air movement in range 0-40 fpm or 0-0.2 m/sec.

Table II. Formulas Relating h_c to Velocity (v)

Formulas	*Activity*	*Ref.*	*Comments*[a]
$h_c = 11.6\text{v}^{0.5}$	Seated	(3)	v, Room air movement by hot wire anemometer
$h_c = (2.7 + 8.7\text{v}^{0.67})$	Reclining	(8)	Air motion lengthwise v, by hot wire anemometer
$h_c = 6.5\text{v}_{tw}{}^{0.39}$	Treadmill	(8)	v_{tw} is speed of treadmill in normal room air
$h_c = 8.6\text{v}_{fw}{}^{0.53}$	Free walking	(8)	v_{fw} is speed of walking in normal room air
$h_c = 8.6\text{v}_{fw}{}^{0.53} + 1.96\text{v}_a{}^{0.86}$	Free walking	See text	v_a is air movement against direction of motion

[a]Measured in meters per second.

The value of h_c varies with atmospheric pressure B and the appropriate factor has been inserted in Eq. (22).

Dry Heat Exchange (DRY)

The total dry heat exchange from the skin surface may be found by combining Eqs. (17) and (22). When written as a loss at sea level,

$$(\text{DRY}) = R + C = F_{\text{cl}}[(h_{\text{r}} + h_{\text{c}})(\bar{T}_{\text{sk}} - T_{\text{a}}) - (\text{ERF})] \qquad (23)$$

In general, the total dry heat exchange can always be separated into two terms–the first describes the Newtonian heat loss from the skin surface $\bar{T}_{sk}$ to a uniform environment at ambient temperature T_a and the second is the effective radiant field.

Operative temperature T_o is defined as the hypothetical temperature of a uniform black enclosure with which man will exchange dry heat at the same rate he would by radiation and convection in the actual nonuniform environment. Thus by definition and from Eq. (11) above

$$\mathrm{DRY} = R + C = h(\bar{T}_{cl} - T_o) \tag{24}$$

$$= F_{cl} h(\bar{T}_{sk} - T_o) \tag{24a}$$

By comparing Eqs. (23) and (24)

$$T_o = T_a + (\mathrm{ERF})/h \tag{25}$$

where

$$h = h_r + h_c \tag{26}$$

In its linear form Eq. (20) may be rewritten as

$$(\mathrm{ERF}) = h_r(\bar{T}_r - T_a) \tag{27}$$

By inserting Eq. (27) in Eq. (25),

$$T_o = (h_r \bar{T}_r + h_c T_a)/(h_r + h_c) \tag{28}$$

which describes the classic definition of operative temperature and states that T_o is an average of $\bar{T}_r$ and T_a, weighted by the transfer coefficients concerned.

For radiation exchange Eq. (25) defines T_o in the *energy* mode, and Eq. (28), in the *temperature* mode.

Evaporative Heat Loss (E)

The evaporation of water vapor to his environment is man's most important key to his ability to survive. The total evaporative heat loss E is the best single physiological index of man's environmental stress. Since the beginning of partitional calorimetry, the observed changes in E have been the quantitative basis for the calibration of A_r/A, and the transfer coefficients for the radiation and convective exchange for the experimental arrangement used, as may be seen in references 1, 16, 27, the studies of Nelson *et al.* (24), and of Colin and Houdas (25).

From the viewpoint of whole-body calorimetry a major advance has occurred recently in the direct measurement of E from the rate of change in body weight. The new advance is a "Potter Bed Balance," described in United States patents 3,224,518 and 3,360,002. This new type of balance is designed without any wearing surfaces, such as knife edges. For the first time it has been possible for physiologists to make continuous measurements of the body weight w during both rest and even heavy exercise (28). Coupled with quite elementary instrumentation, a minute by minute evaluation of body weight w is now possible. E is determined from the rate of weight loss w by the relation

$$E = 60 \cdot w \cdot \lambda / A_D \tag{29}$$

where $\dot{w}$* is in gm/min and the latent heat λ is equal to 0.68 W · hr/gm.

The total evaporative loss E, as mentioned above, may be attributed to three sources–(1) the respired vapor loss E_{res}, (2) the diffusion of water vapor through the skin layers E_{diff} and (3) the evaporation E_{rsw} of sweat secreted on the skin surface necessary for temperature regulation. The net heat loss from the skin surface is

$$E_{sk} = E - E_{res} = E_{diff} + E_{rsw} \tag{30}$$

At this point there are two apparently similar but very different questions to be asked–(a) What is the maximum possible heat loss by sweating that can be evaporated from a normal skin surface for any given combination of environmental factors? (b) If the skin surface is assumed to be 100% wet, what would be the maximum evaporative heat loss possible to the environment based on the meteorology involved?

The first question (a) involves two factors–What is the maximum capacity of the body to secrete sweat? What is the maximum percent of the total skin surface that can be made wet by sweating? For the first, Kerslake and Brebner (29) has demonstrated a maximum of 30gm/min for short periods and a value of 20 gm/min has been proposed by those interested in extended performance in the heat (30). As for the second factor, there is currently no clear evidence to this date to show that the body can become 100% wet by normal sweating even at high humidities.

Using methods to be described later, preliminary analysis of published data (31, 32) would show for unclothed sedentary subjects the maximum skin wettedness falls in range 80-100% at the temperature and humidity when regulation fails and mean body temperature rises. With exercising unclothed

*For heavy exercise $\dot{w}$ must be corrected for CO_2 loss over O_2 gain by minus $\Delta = 0.00154 \cdot M \cdot A_D$.

subject (28) the probable range is 90-100%. With normally clothed subjects, the flow of sweat into the clothing layers can easily reduce this effectiveness factor down to 50%. As pointed out by Kerslake (33) in an analytic study on cylinders, air turbulence tends to improve the wetting efficiency of sweating.

In the desert with its low ambient vapor pressures and by wearing loosely fitting clothing, there is likely no serious loss of efficiency in sweating on the skin surface. The maximum may rise again to the 90-100% range. This speculative question remains a possible point for future experimentation. For the present analysis we will assume this coverage factor to be 100% effective.

The importance of the second question (b), regarding the maximum possible loss from an assumed 100% wet skin surface, was first made by Belding and Hatch (30) when they proposed their Heat Stress Index (HSI) in 1956. In terms of the environment, the maximum heat loss E_{max} from a saturated skin surface is given by

$$E_{max} = h_e(P_{sk} - P_{dew})F_{pcl} \quad \text{W/m}^2 \tag{31}$$

where h_e is the evaporative heat transfer coefficient from the skin surface in $\text{W/(m}^2 \cdot \text{mm Hg)}$.

Engineers (34) and meteorologists (35) have long been familiar with the analogous relationship that exists between the processes of evaporation and convection, but only recently has this been brought into sharp focus for physiologists (5, 6). Mass and convective heat-transfer theory has shown that the following ratio is valid

$$h_e/h_c = 2.2(760/B) = (\text{LR}) \quad {}^\circ\text{C/mmHg} \tag{32}$$

This ratio is known as the Lewis Relation (LR). Combining (31) and (32)

$$E_{max} = (\text{LR})h_c(P_{sk} - P_{dew})F_{pcl} \tag{33}$$

It is important to note that, as the Lewis Relation increases with altitude, the value for h_c lowers with altitude. Thus at 10,000 ft altitude there would be a 20% net increase in E_{max} if other factors were unchanged.

The diffusion of water vapor from the skin surface can be an important factor in dry climates. Brebner *et al.* (36) have shown a close correlation between the water loss by skin diffusion and the vapor pressure gradient to the environment when no regulatory sweating is present. Using their data, it can be shown that, when $E_{rsw} = 0$, E_{diff} is approximately 6% of the associated E_{max}. This percent value may be considered as a maximum and may drop as the skin layers dry on exposure to low humidity. For the present analysis we will assume any skin surface area not sweating will have a diffusion loss of 0.06 E_{max}. Finally, it should be recognized that E_{diff} and E_{rsw} never occur on the *same* skin area at the *same* time. Thus E_{diff} for the whole body surface is at its maximum when sweat secretion (E_{rsw}) is zero, and E_{diff} will approach zero when $E_{rsw} = E_{max}$ or when surface is 100% wet.

The ratio of E_{sk}/E_{max} may be used as a measure of the total skin wettedness w caused by both diffusion and sweating. The wettedness caused by the regulatory sweat alone is w_{rsw} or E_{rsw}/E_{max}. Thus the total skin evaporative heat loss may now be described by the relation:

$$E_{sk} = (0.06 + 0.94 w_{rsw}) E_{max} \tag{34}$$

When w_{rsw} and w equals unity, $E_{sk} = E_{max}$. If E_{sk} is directly measurable by $(E - E_{res})$, w_{rsw} may be calculated by Eq. (34). Since $E_{rsk} = w_{rsw} E_{max}$, then the actual value E_{diff} follows from Eq. (30).

When $E_{rsw} > E_{max}$, sweat rolls off the skin surface and its cooling value will be lost for purposes of heat regulation. This lost cooling as a minimum is given by

$$E_{drip} = E_{rsw} - E_{max} \tag{35}$$

The reader should note that any evaporative rate E may be converted to a rate of weight loss in gm/min for the total body by multiplying by the factor $(A_D/42)$.

Mean Body Temperature, $\overline{T}_b$

The change in mean body temperature at any time t is related to the storage S by

$$\Delta\overline{T}_b/\Delta t B t = (S \cdot A_D/(0.97 \cdot w) \tag{36}$$

where 0.97 is the body's specific heat in (W · hr)/(kg · °C).

If the mean body temperature $\overline{T}_b$ can be determined at the start of the experiment by a weighted average of $\overline{T}_{sk}$ and T_{re}, using the Burton and Edholm (10) or Hardy *et al.* (37) ratios of 1 : 2 for the cold or 1 : 4 for warmth, it is possible to predict $\overline{T}_b$ at any time t by

$$\overline{T}_b = (a\overline{T}_s + bT_{re})/(a + b) + \Sigma_o^t(\Delta\overline{T}_b/\Delta t)dt \tag{37}$$

where a : b is the selected weighting ratio initially.

Skin Conductance (K), or Skin Blood Flow SKBF

The transfer of heat from the central core to the skin surface is a function of the specific conductivity of the skin layer, the depth of the thermal gradient, and the blood flow (SKBF) (current and countercurrent). Partitional calorimetry can describe only the average picture. Skin conductance K at any time is given by the relation

$$K = (E_{sk} + R + C)/(T_{cr} - T_{sk}) \qquad W/(m^2 \cdot {}^\circ C) \qquad (38)$$

The minimum value of K occurs when $E_{rsw} = 0$ and while the body is in a state of body cooling (i.e., $S<0$). An average minimum value (31) for man is approximately 5.3 $W/(m^2 \cdot {}^\circ C)$. Since the specific heat of blood is 1.163 $W \cdot hr/(kg \cdot {}^\circ C)$ and 1 kg $\cong$ 1 liter, an estimate of skin blood flow (SKBF) from K as measured by Eq. (38) would be

$$(SKBF) = (K - 5.3)/1.163 \qquad \text{in liters}/(m^2 \cdot hr) \qquad (39)$$

Applications of Partitional Calorimetry

In any calorimetric experiment or application it is essential to establish values for, measure directly or control the *independent* variables listed above, namely M, W, T_a, T_{dew} (or humidity), h_c or air movement, h_r (or $\overline{\overline{T}}_w$ or ERF), I_{clo}, B and time of exposure t. The *dependent* variables, which include all the physiological variables except M and W, are $\overline{T}_{sk}$, T_{cr} (by T_{re}, T_{es}, and/or T_m) E, and $(SKBF)$.

There are three principal uses of partitional calorimetry—(1) quantitative measurement of all avenues of energy exchange; (2) the evaluation of the transfer coefficients involved in the exchange of heat between the body and the environment; and (3) the establishment of quantitative relationships between the four physiologically dependent variables during rest and exercise.

One of the first and simplest uses of modern calorimetry in the desert was accomplished (perhaps unknowingly) by Gosselin (38) during his 1942 studies on rates of sweating. In his study (see Fig. 3-8 in reference 38), he compared the average sweating rate of soldiers while sitting in the sun and shade at the same ambient temperature and reported an average change of 165 gm/hr. As can be seen from our heat balance equations during relative thermal equilibirum, the increase in sweating ΔE is caused by the net radiant load on the body, i.e., $(ERF) \times F_{cl}$. By assuming that the average body surface area was 1.8 m^2, the subjects were wearing 1 clo, the ambient air movement was about 5 mph, F_{cl} is

0.4, $\alpha = 0.7$, and the average shape factor $F_{m-s} = 0.33$ for all solar radiation, the following deductions are possible:

Net heat load on body (ERF × F_{cl}) is 165 gm/hr or 165 × 0.7/1.8 or 64 W/m^2.

Net heat load on outer clothing surface (ERF) is 64/0.4 or 160 W/m^2.

Net incident irradiation to body surface is 160/0.7 or 230 W/m^2.

Total incident irradiation from the sun, including the direct, the diffuse, and the albedo, is 230/0.33 or 700 W/m^2.

The above values agree roughly in magnitude with the direct measurements of Roller and Goldman reported above. In these experiments Gosselin was in reality using man as a direct reading 4π radiometer of solar radiation.

The evaluation of transfer coefficients for any given experimental arrangement is essentially a calibration procedure. It is best done, as pointed out above, when $\Delta \bar{T}_b/\Delta t$ *or* S is zero, and either by varying the radiation exchange while the convection exchange is held constant or vice versa. The best test to find when S is zero is the use of a "linearity criterion" (27) between two major groups of variables. A classic example (39) is shown in Fig. 2, in which some of the early data of Winslow (26) and more recent data of Colin and Houdas (25) are used for the evaluation of the combined heat transfer coefficient ($h_r + h_c$) or (h).

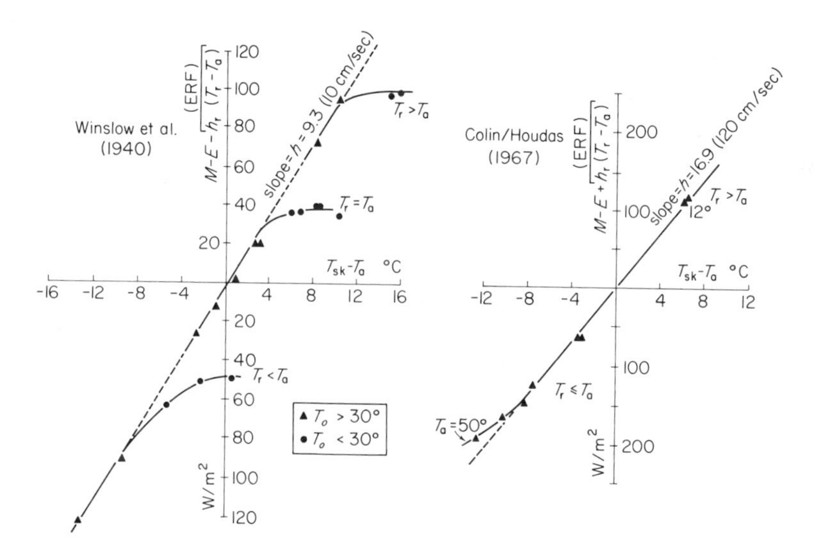

FIG. 2. Use of partitional calorimetry to calibrate combined heat transfer coefficients in a test environment (25, 47). The sum of the metabolic rate and effective radiant field less the evaporative heat loss is plotted against the gradient from skin to ambient air. The slope is the combined transfer coefficient. Deviation from linearity indicates presence of body heating ($S > 0$) or cooling ($S < 0$).

Calibration type experiments in partitional calorimetry such as those in Fig. 2 are for the most part accomplished in the operative temperature range from 30-40° C, where the regulatory sweating by the resting subject has its greatest variability and where it can be measured with best accuracy.

Figure 3 is an illustration (40) of minute-by-minute partitional calorimetry on a subject working to exhaustion at 90% maximum work out. The initial value for $\bar{T}_b$ was based on the Hardy-Dubois weighting for $\bar{T}_{sk} : T_{re}$ or 1 : 4 and the $\Delta\bar{T}_b$ for each succeeding minute was calculated from S and added to the initial $\bar{T}_b$. During the exercise phase it was found that statistically

$$\Delta\bar{T}_b = 0.21\Delta\bar{T}_{sk} + 0.51\Delta T_{es} + 0.28\Delta T_m \qquad (40)$$

with a confidence factor of 0.96.

The whole body studies of Benzinger (41) in a rapid acting gradient calorimetry initiated early during the past decade a tremendous interest on the part of physiologists in the true relationship between regulatory sweating E_{rsw} or E and the associated internal body and skin temperatures during both rest and exercise. Such studies can be accomplished quantitatively in an exploratory sense without calorimetry by the use of sweat capsules but, if these relationships are meaningful quantitatively in terms of the whole body energy exchange

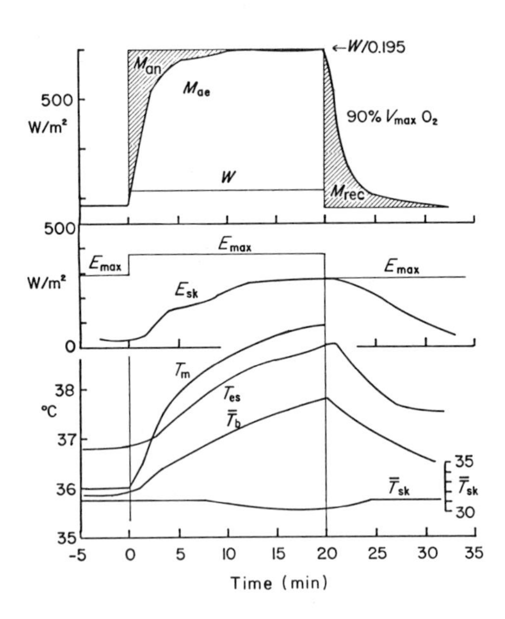

FIG. 3. A minute-by-minute heat partition during maximal exercise to exhaustion. The ambient air temperature was 20°C.

during temperature regulation, they must be performed under conditions where the associated partitional calorimetry is known quantitatively. Such a classic plot (42) is shown in Fig. 4 for sedentary subjects showing how the observed E_{rsw} varies with skin and tympanic temperature. From this relation Hardy and Stolwijk derived their control equation for sweating while sedentary. From Fig. 4 this may be stated as follows.

$$E_{rsw} = 70.0(T_{ty} - 36.6)(\bar{T}_{sk} - 33.5) \qquad (41)$$

in W/m^2 with a confidence factor of 0.9.

In applying Eq. (41) if either term in parenthesis is zero or negative, then $E_{rsw} = 0$. This "double-delta" rate control relationship, although experimentally valid for sedentary subjects, obviously does not satisfy observed E_{rsw} vs ($\bar{T}_{sk}$, T_{re} or T_{es}) relationships during exercise, since $\bar{T}_{sk}$ can easily fall below 33.5 during heavy sweating at lower (~ 20°) ambient temperature.

By using the same threshold as above for T_{es} of 36.6, Saltin *et al.* (28) observed the following relationships during exercise

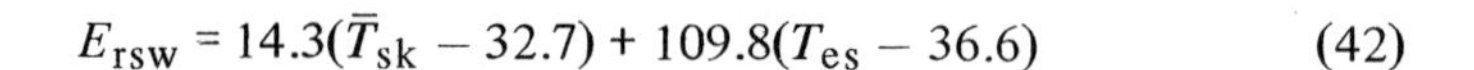

$$E_{rsw} = 14.3(\bar{T}_{sk} - 32.7) + 109.8(T_{es} - 36.6) \qquad (42)$$

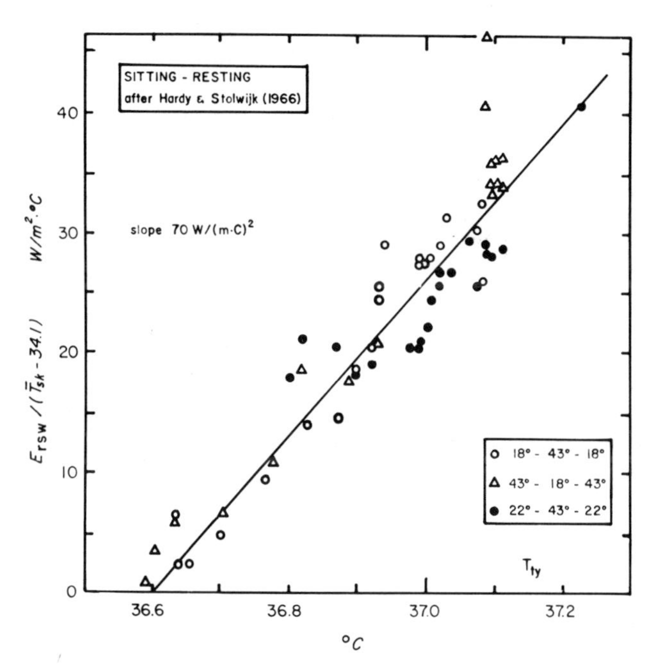

FIG. 4. Observation of E_{rsw} in relation to mean skin and tympanic temperatures for a sedentary unclothed subject. [After Hardy and Stolwijk (42).]

where the confidence factor is 0.8. From partitional calorimetry data in their South African test chamber, Wyndham and Adkins (43) have observed that the general form for E_{sk} may be

$$E_{rsw} = 430(T_{re} - 36.5)[0.1 + 0.455(\bar{T}_{sk} - 33.3)/3.7)] \tag{43}$$

More recently Nadel, Bullard, and Stolwijk have presented data (44) that may validate the following type relationship

$$E_{rsw} = [17.5(\bar{T}_{sk} - 34.1) + 175(T_{es} - 36.6)]\, e^{(T_{sk} - 34.1)/10} \tag{44}$$

All three experimental trends do show the importance of $\bar{T}_{es}$ or T_{re} for setting the drive for secretion of sweat and of $\bar{T}_{sk}$ in modifying this drive peripherally. More important for the purpose here, they show the type of fundamental information that may be derived by partitional calorimetry.

A final example (40) of the use of calorimetry is a plot of E_{rsw} vs $\bar{T}_b$ during exercise and recovery from a 90% maximum work run (Fig. 5). The data represented are an average of observations on four subjects. $\bar{T}_b$ has been calculated by the integrating $\Delta\bar{T}_b$ for each minute of the experiment using Eq. (37). This figure shows the importance of mean body temperature $\bar{T}_b$ as related to sweating during both exercise and recovery.

Use of a Simple Model of Human Temperature Regulation to Predict Partitional Calorimetry

If one examines the partitional calorimetry for the passive state, one will recognize that, if there is a known relationship between (SKBF) and E_{rsw} and the body temperature $\bar{T}_{sk}$, T_{re}, and T_{es}, it will then be possible to develop an analytic model that can predict the four dependent variables above from the environmental factors concerned. This suggestion was originally made by Stolwijk and Hardy (45) and has since been the fundamental basis of many models of temperature regulation, now used by NASA and in engineering literature.

As an illustration of the Stolwijk-Hardy principle, let us consider a simple 2-node model of the human body—a core at temperature T_{cr} and a thin skin shell at $\bar{T}_{sk}$, which model is similar to the electric analog originally proposed by MacDonald and Wyndham (46) in 1950. Let us give our 2-node model the characteristic of a standard-size man (70 kg-1.80 m^2), and assume that he is at thermal physiological neutrality at the start of exposure and that he exchanges heat with his environment in accordance with the passive state equations listed above. If (HFCR) is the net heat flow by metabolic heat to the core surface and

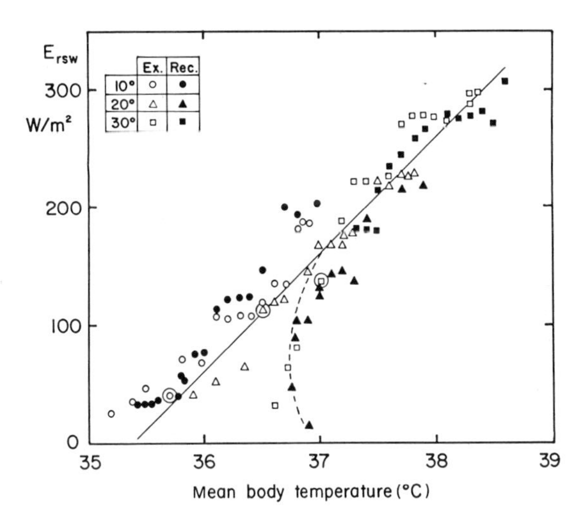

FIG. 5. Comparison of regulatory sweating (E_{rsw}) with mean body temperature calculated each minute from observed $\Delta\bar{T}_b/\Delta t$ or S by the heat balance equation. The exercise was bicycling at 60 rpm and 90% $\dot{V}maxO_2$. The dotted curve shows trend for initial and final 30° points as we:l as 20° recovery.

if (HFSK) is the heat flow from the skin surface to the environment, then the dual passive-state heat-balance equations are

$$(\text{HFCR}) = M_{net} - (T_{cr} - \bar{T}_{sk})[5.3 + 1.163 \cdot (\text{SKBF})] \tag{45}$$

$$(\text{HFSK}) = (T_{cr} - \bar{T}_{sk})[5.3 + 1.163(\text{SKBF})] - E_{sk} - (\text{DRY}) \tag{46}$$

However,

$$S = (\text{HFCR}) + (\text{HFSK}) \tag{47}$$

and is given by the basic equation (6).

If we assign 7.0 kg and 63.0 kg to be the respective masses of the skin shell and body core then

$$\Delta\bar{T}_{sk} = (\text{HFSK})(1.8)/(0.97)(7.0) \tag{48}$$

and

$$\Delta T_{cr} = (\text{HFCR})(1.8)/(0.97)(63.0) \tag{49}$$

where 0.97 W · hr/(kg · °C) is the body specific heat.

By integrating (48) and (49) each minute from the initial neutral values $\bar{T}_{sk} = 34.1$ and $T_{cr} = 36.6$ at $t = 0$ over any time t, and by relating at each successive minute the new $\bar{T}_{sk}$ and new T_{cr} to the skin blood flow (SKBF) and E_{rsw} generated, it will be possible to predict after any time t the probable value of $\bar{T}_{sk}$, T_{cr}, E_{rsw}, and (*SKBF*) for any environmental condition and activity described by the independent variables above. An annoted FORTRAN program for such a simple model is given in the appendix. The reader should have no trouble following the FORTRAN definitions as they are very similar to the algebraic terms used throughout this chapter. Figure 6 shows how this 2-node model calls for sweating (E_{rsw}) and skin blood flow (*SKBF*) for various values of T_{sk},T_{cr}, and metabolic rates (bicycling). The curves for 1-6 "mets" represent a 1-hr exposure; 9 mets, 45-min exposure; and 12 mets, 15-min exposure.

A simple model of human temperature regulation, such as is described above can be used to explore human response over a wide range of environmental conditions when tied in with the detailed heat balance equations defined above. Examples of the type of predictive partitional calorimetry possible are given as follows.

1. The heat balance predicted for a normally clothed subject when exposed for 1 hr at various ambient air temperatures and two humidities is shown in Fig. 7. The model shows that humidity does not affect skin and core temperature, the rate of evaporative cooling or skin blood flow until the surface becomes 100% wet. The data of Winslow *et al.* (47, 48) confirmed this observation.

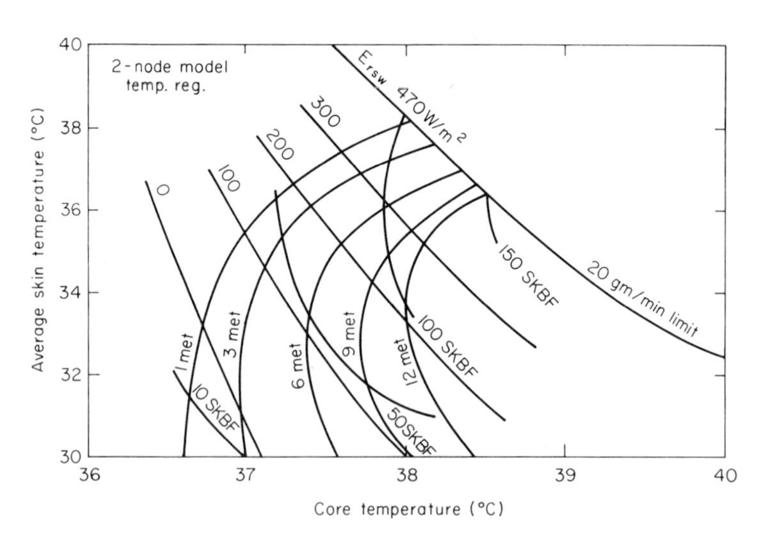

FIG. 6. Loci of constant metabolic rate (met), constant regulatory sweating (E_{rsw}), and constant skin-blood flow (SKBF) in terms of $\bar{T}_{sk}$ and T_{cr}, as predicted by the 2-node model described in the appendix. 1 met = 58.2 W/m².

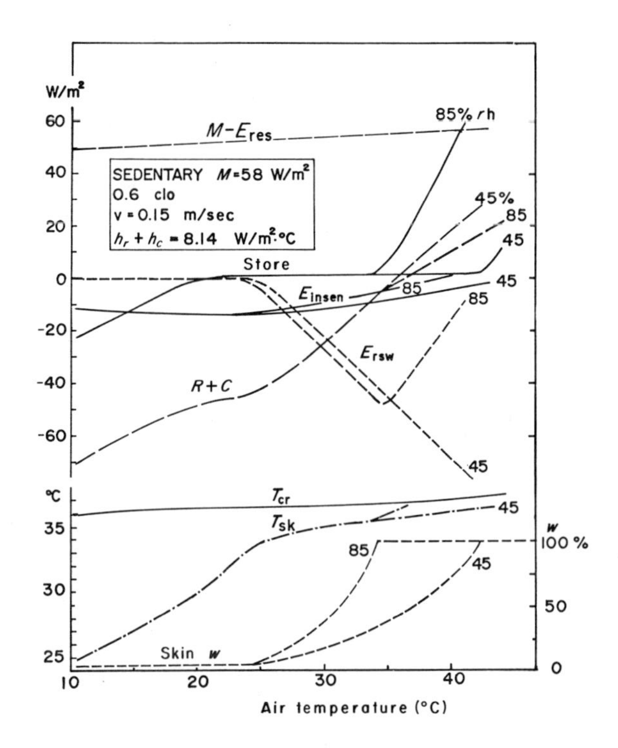

FIG. 7. Predicted heat balance (partitional calorimetry) after 1-hr exposure to various dry heat temperatures and humidities.

2. Effect of clothing on setting the upper and lower limits of regulatory sweating is shown in Fig. 8. The fact that clothing insulation changes very little the upper limit of regulation by sweating also holds true for higher but constant levels of activity. The upper limit lowers approximately 1° C (on 100% *rh* line) for each *met* of activity (i.e., each multiple of resting metabolic rate). The lower limit for thermal neutrality without regulatory sweating varies very considerably with exercise toward freezing temperatures—so much so that other factors, such as selective cooling of the extremities, complicate the theoretical picture described by the model.

3. Use of clothing as protection during a walk in the desert in the early afternoon is shown in Fig. 9. This figure demonstrates a fact well known to those working in the sun and in the dry heat. For little or no clothing the physiological limit is set by the ability to produce sweat. As the insulation of clothing and its protection against dry heat increases, the microclimate next to the skin surface eventually results in a saturated skin surface and a general failure of the cooling process by sweating. For the case illustrated the skin blood flow is

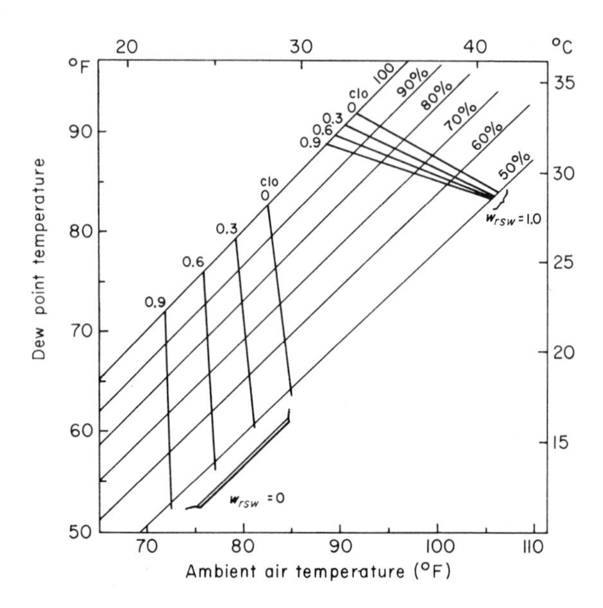

FIG. 8. Effect of clothing on setting the upper and lower limits of regulatory sweating for a sedentary subject: 30 fpm air velocity.

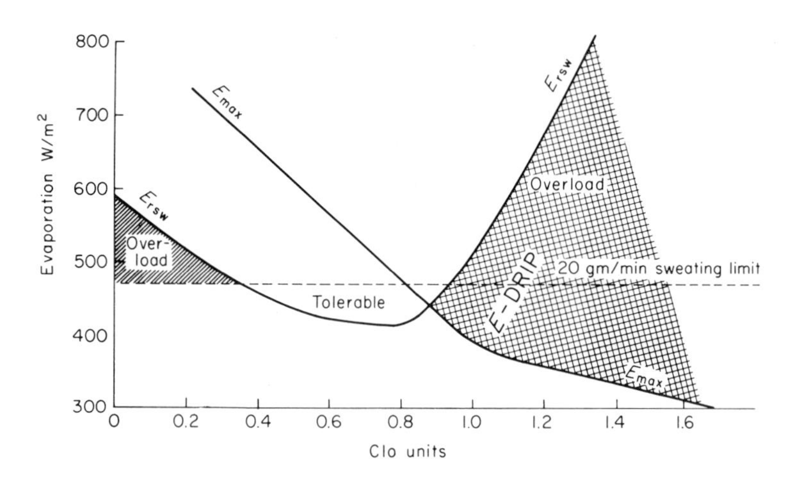

FIG. 9. The relation of regulatory sweating to clothing insulation while walking in the desert. M = 350 W/m² (6 met), walking 3 mph, in a 10 mph wind, at 5000 ft altitude. $T_a = 43°C$, $T_{dew} = 11.3°C$, $T_r = 81°C$, $T_o = 53°C$, *and* $ERF = 200\ W/m^2$.

above 100 liters/$(m^2 \cdot hr)$ for the "tolerable" zone and for this reason alone our "subject" would be still under considerable strain.

Summary

1. For the successful use of partitional calorimetry in the desert the following minimal basic measurements are required—the rate of O_2 consumption, the ambient temperature, the humidity (i.e., T_{dew}, rh, T_{wet}, or millimeters of mercury of ambient water vapor pressure), the mean skin temperature, an internal measure of body temperature (T_{re}, T_{es}, or T_{ty}), and the change in body weight over the exposure period.

2. When radiation is present, a measurement of the ERF is necessary with the use of a globe thermometer or some other type of 4π radiometry. The absorptance of exposed surfaces to solar or other thermal radiation present is the second necessary factor.

3. When clothing is used, either the clothing surface temperature must be measured or the clothing insulation be known from previous measurements on a heated, man-sized manikin.

4. The most important single coefficient to be evaluated either by the calorimetric measurement or by direct methods is the convective heat transfer coefficient. This coefficient, in addition to the dry heat exchange, affects the thermal efficiency of clothing both for transfer of heat as well as of water vapor, and the magnitude of the maximum evaporative heat capacity to the environment.

5. A simple model of human temperature regulation may be used to predict partitional calorimetry under widely varying environmental conditions.

References

1. Winslow, C.-E. A., Herrington, L. P., and Gagge, A. P. (1936). A new method of partitional calorimetry. *Amer. J. Physiol.* **116**, 641.
2. Hardy, J. D., and Dubois, E. F. (1937). Regulation of heat loss from the human body. *Proc. Nat. Acad. Sci. U.S.* **23**, 624.
3. Fanger, P. O. (1967). Calculation of thermal comfort. *ASHRAE Trans.* **73**, 4.1.
4. Fanger, P. O. (1970). "Thermal Comfort: Analysis and Applications in Environmental Engineering." Danish Tech. Press, Copenhagen.
5. Rapp, G. M. (1971). Convective mass transfer and the coefficient of evaporative heat loss from human skin. *In* "Physiological and Behavioral Temperature Regulation" (J. D. Hardy *et al.*, eds.), Chapter 6, p. 55. Thomas, Springfield, Illinois.
6. Nishi, Y., and Ibamoto, K. (1969). Model skin temperature—an index of thermal sensation in cold, warm and humid environments. *ASHRAE Trans.* **75**, 94.

7. Nishi, Y., and Gagge, A. P. (1970). Moisture permeation of clothing–A factor governing thermal equilibrium and comfort. *ASHRAE Trans.* **76**, 137.
8. Nishi, Y., and Gagge, A. P. (1970). Direct evaluation of convective heat transfer coefficient by naphthalene sublimation. *J. Appl. Physiol.* **29**, 830.
9. Hardy, J. D. (1949). Heat transfer. *In* "Physiology of Heat Regulation and the Science of Clothing" (L. H. Newburgh, ed.), Chapter 9, p. 78. Saunders, Philadelphia, Pennsylvania.
10. Burton, A. C., and Edholm, O. G. (1955). "Man in a Cold Environment." Arnold, London.
11. Dubois, D., and Dubois, E. F. (1916). A formula to estimate approximate surface area if height and weight are known. *Arch. Intern. Med.* **17**, 863.
12. Chambers, A. B. (1970). A psychrometric chart for physiological research. *J. Appl. Physiol.* **29**, 406.
13. Smithsonian Meteorological Tables. (W. E. Forsythe, ed.) (1954). Smithsonian Inst., Washington, D.C.
14. Spofford, W. A. (1969). Air conditioning, cooling and heating capacity and air flow calculated by computer. (See equation 1.) *ASHRAE Trans.* **75**, 207.
15. Raber, B. F., and Hutchinson, F. W. (1947). "Panel Heating and Cooling Analysis." Wiley, New York.
16. Gagge, A. P., Rapp, G. M., and Hardy, J. D. (1967). The effective radiant field and operative temperature necessary for comfort with radiant heating. *ASHRAE Trans.* **73**, 2.1.
17. Hardy, J. D., and Dubois, E. F. (1938). The technique of measuring radiation and convection. *J. Nutr.* **15**, 461.
18. McCutchan, J. W., and Taylor, C. L. (1951). Respiratory heat exchange with varying temperatures and humidities of inspired air. *J. Appl. Physiol.* **4**, 121.
19. Hardy, J. D., Hammel, H. T., and Murgatroyd, D. (1956). Spectral transmittance and reflectance of excised human skin. *J. Appl. Physiol.* **9**, 257.
20. Roller, W. L., and Goldman, R. F. (1968). Prediction of solar heat load on man. *J. Appl. Physiol.* **24**, 717.
21. Richards, C. H., Stoll, A. M., and Hardy, J. D. (1951). The pan-radiometer: An absolute measuring instrument for environmental radiation. *Rev. Sci. Instrum.* **22**, 925.
22. Sutton, D. J., and McNall, P. E., Jr. (1954). A two-sphere radiometer. *ASHVE Trans.* **60**, 297.
23. Gagge, A. P., Graichen, H., Stolwijk, J. A. J., Rapp, G. M., and Hardy, J. D. (1968). The R-meter. A globe shaped skin colored sensor of radiant field, operative temperature and the associated environmental variables that affect thermal comfort. *ASHRAE J.* **10**, 77.
24. Nelson, N., Eichna, L. W., Horvath, S. N., Shelley, W. B., and Hatch, T. F. (1947). Thermal exchange of man at high temperature. *Amer. J. Physiol.* **141**, 626.
25. Colin, J., and Houdas, Y. (1967). Experimental determination of coefficients of heat exchange by convection of the human. *J. Appl. Physiol.* **22**, 31.
26. Winslow, C. -E. A., and Gagge, A. P. (1941). Influence of physical work on physiological reactions to the thermal environment. *Amer. J. Physiol.* **134**, 664.
27. Gagge, A. P. (1936). The linearity criterion as applied to partitional calorimetry. *Amer. J. Physiol.* **116**, 656.
28. Saltin, B., Gagge, A. P., and Stolwijk, J. A. J. (1970). Body temperatures and sweating during thermal transients caused by exercise. *J. Appl. Physiol.* **23**, 318.
29. Kerslake, D. McK., and Brebner, D. F. (1971). Maximum sweating at rest. *In* "Physiological and Behavioral Temperature Regulation" (J. D. Hardy *et al.*, eds.),

Chapter 10, p. 139. Thomas, Springfield, Illinois.
30. Belding, H. S., and Hatch, T. F. (1956). Index for evaluating heat stress in terms of resulting physiological strain. *ASHRAE Trans.* **62**, 213.
31. Stolwijk, J. A. J., and Hardy, J. D. (1966). Partitional calorimetric studies of responses of man to thermal transients. *J. Appl. Physiol.* **21**, 967.
32. Gagge, A. P. (1937). A new physiological variable associated with sensible and insensible perspiration. *Amer. J. Physiol.* **120**, 277.
33. Kerslake, D. McK. (1963). Errors arising from the use of mean heat exchange coefficients in the calculation of the heat exchanges of a cylindrical body in a transverse wind. *Temp. Its Meas. in Sci. Ind., Proc. Symp., 4th, 1961.* Vol. 3, Part 3, p. 183.
34. Missenard, A. (1934). Théories et lois nouvelles de l'évaporation. *Chal. Ind.* **15**, 129.
35. Buettner, K. J. K. (1934). Die Wärmeübertragung durch Leitung and Konvektion, verdustung and strahlung in Bioclimatologie und Meteorologie. *Veroef. Preuss. Meteorol. Inst.* **10**, No. 5,
36. Brebner, D. F., Kerslake, D. McK., and Waddell, J. L. (1956). The diffusion of water vapor through human skin. *J. Physiol. (London)* **132**, 225.
37. Hardy, J. D., Milhorat, A. T. , and Dubois, E. F. (1941). Basal metabolism and heat loss of young women at temperatures from 22°C and 35°C. *J. Nutr.* **21**, 353.
38. Gosselin, R. E. (1947). Rates of sweating in the desert. *In* "Physiology of Man in the Desert" (E. F. Adolph, ed.), Chapter IV, p. 77. Wiley (Interscience), New York.
39. Gagge, A. P. Effective radiant flux, an independent variable that describes thermal radiation on man. *In* "Physiological and Behavioral Temperature Regulation" (J. D. Hardy *et al.*, eds.), Chapter 4, p. 34. Thomas, Springfield, Illinois.
40. Gagge, A. P., and Saltin, B. (1971). Prediction of human body temperature during maximal exercise from skin, rectal, esophageal and muscle temperatures. *Proc. Int. Congr. Physiol. Sci., 25th, 1971,* IX, 192.
41. Benzinger, T. (1959). On the physical heat regulation and the sense of temperature in man. *Proc. Nat. Acad. Sci. U.S.* **45**, 645.
42. Hardy, J. D., and Stolwijk, J. A. J. (1966). Partitional calorimetric studies of man during exposure to thermal transients. *J. Appl. Physiol.* **21**, 1799.
43. Wyndham, C. H., and Atkins, A. R. (1968). A physiological scheme and mathematical model of temperature regulation in man. *Pfluegers Arch.* **303**, 14.
44. Nadel, E. R., Bullard, R. W., and Stolwijk, J. A. J. (1971). The importance of skin temperature in the regulation of sweating. *J. Appl. Physiol.* **31**, 828.
45. Stolwijk, J. A. J., and Hardy, J. D. (1966). Temperature regulation in man—a theoretical study. *Pfluegers Arch. Gesamte Physiol. Menschen Tiere* **291**, 129.
46. MacDonald, D. K. C., and Wyndham, C. H. (1950). Heat transfer in man. *J. Appl. Physiol.* **3**, 342.
47. Winslow, C. -E. A., Gagge, A. P., and Herrington, L. P. (1940). Heat exchange and regulation in radiant environments above and below air temperature. *Amer. J. Physiol.* **131**, 79-92.
48. Winslow, C. -E. A., Herrington, L.,P., and Gagge, A. P. (1938). The reaction of the clothed human body to variations in atmospheric humidity. *Amer. J. Physiol.* **124**, 692.
49. For a detailed program, see Appendix, Gagge, A. P., Stolwijk, J. A. J., and Nishi, Y. (1971). *ASHRAE Trans. (I)* **77**, 247.

APPENDIX

AN ANNOTATED FORTRAN PROGRAM FOR A SIMPLE CORE-SHELL MODEL OF HUMAN TEMPERATURE REGULATION (49)

```
C     THE INDEPENDENT VARIABLES FOR TYPE OF ACTIVITY
C     ARE DEFINED IN INITIAL READ STATEMENTS BY:
C        RM, WK, CLO, CHR, CHC, TA, TDEW, ERF, B, AND TIMEX
C     "TIMEX" IS EXPOSURE TIME TO ENVIRONMENT IN HOURS.
C     UNITS ARE WATTS SQ. METER AND DEG. C
C
C     STANDARD MAN:     70 KG AND 1.8 SQ. M DUBOIS AREA.
C                       SKIN MASS - 7.0 KG
C                       CORE MASS - 63.0 KG
C
C     INITIAL CONDITIONS – PHYSIOLOGICAL THERMAL NEUTRALITY
C       TSK = 34.1   TCR = 36.6   ERSW = 0  SKBF = 6.3   TIME = 0
C     RMNET, ESK AND DRY ARE DEFINED
C
C     HEAT BALANCE EQUATIONS FOR PASSIVE STATE
C
       600    HFSK = (TCR-TSK)*(5.3 + 1.163*SKBF) - ESK - DRY
              HFCR = RMNET - (TCR-TSK)*(5.3 + 1.163*SKBF)
C     THERMAL CAPACITY OF SKIN SHELL (TCSK) IN W·HR/KG·C
              TCSK = 0.97*7.0
C     THERMAL CAPACITY OF CORE SHELL (TCCR)
              TCCR = 0.97*63.0
C     CHANGE IN SKIN (DTSK) AND CORE (DTCR) TEMPERATURE
              DTSK = (HFSK*1.8)/TCSK
              DTCR = (HFCR*1.8)/TCCR
              DTIM = 1/60
              TIME = TIME + DTIM
              TSK = TSK + DTSK*DTIM
              TCR = TCR + DTCR*DTIM
C
C     CONTROL SYSTEM
C
C     DEFINITION OF SIGNALS FOR THERMOREGULATION
C
C     FROM SKIN
              SKSIG = (TSK -34.1)
              IF (SKSIG) 900, 900, 901
       900    COLDS = – SKSIG
              WARMS = 0.0
              GO TO 902
       901    COLDS = 0.0
              WARMS = SKSIG
```

```
C      FROM CORE
  902      CRSIG = (TCR -36.6)
           IF (CRSIG) 800, 800, 801
  800      COLDC = - CRSIG
           WARMC = 0.0
           GO TO 802
  801      WARMC = CRSIG
           COLDC = 0.0
C
C      CONTROL OF SKIN BLOOD FLOW
C
  802      STRIC = 0.5*COLDS
           DILAT = 75.*WARMC
           SKBF = (6.3 + DILAT)/(1. + STRIC)
C
C      CONTROL OF EVAPORATIVE REGULATION
           IF (RM -70.) 401, 402, 402
C      WHILE SEDENTARY
  401      ERSW = 70.*WARMC*WARMS
           GO TO 403
C      DURING EXERCISE
  402      ERSW = (17.5*SKSIG + 175.*CRSIG)*EXP(SKSIG/10).
           IF (ERSW) 404, 405, 405
C
  403      IF (TIME-TIMEX) 600, 601, 601
  404      ERSW = 0
  405      CONTINUE
  601      CONTINUE
C
C      AT END OF EXPOSURE "TIMEX", MODEL WILL PREDICT: TSK, TCR,
         ERSW, SKBF
C      AS WELL AS ALL FACTORS CONCERNED IN HEAT BALANCE EQUATION
```

IV

Sweat Mechanisms in Heat

F. N. CRAIG

Sweat Mechanisms in Heat

Dill's interest in sweat embraces both the quantity and the chemical composition. This chapter will ignore the latter with one exception related to his early observation at Boulder City that the concentration of lactate was greater in sweat than in plasma. How the sweat glands could extract lactate from the plasma and concentrate it is puzzling. An explanation has been found (1) in a study of isolated human eccrine sweat glands. Enough lactate was produced in the metabolism of the isolated glands to account for the amount in thermal sweat.

Sweating is important in the regulation of body temperature. The theoretical framework has been reviewed by von Euler (2) and Hardy (3). The rate of sweating, however, depends so much on the body temperature that the system for the regulation of sweating requires consideration apart from the mechanism of regulating body temperature. When a man is exposed to heat, sweating does not begin until the body temperature exceeds a certain value called the set point. Then sweating increases in proportion to the elevation in body temperature above the set point, and the slope of the relationship between sweat rate and body temperature represents the gain of the system. The increase continues until the rate reaches a maximum beyond which further increases in body temperature are ineffective. The sweat-regulating system thus may be defined operationally in terms of the set point, gain, and maximum. Observation of how these quantities can be manipulated should improve our understanding of sweat mechanisms in heat.

Within the system for regulating sweating, a number of subsystems can be distinguished. Each of these may have its own set point, gain, and maximum. Hensel (4) has described one such subsystem consisting of a spot 1 or 2 mm in diameter on the skin of the nose of the cat and a single fiber in the infraorbital nerve. Heat applied to the spot increases the frequency of impulses passing in the nerve in direct proportion to the increase in temperature of the skin. The receptor area is highly specific for heat; it does not respond to cold or to mechanical pressure.

Evidence shows that the sweat glands are controlled by centers in the hypothalamus, and the activity of the centers is influenced by the temperature of the blood perfusing them. This depends, in turn, on the temperature of the blood from central and peripheral regions. Another factor is the neural input from central and peripheral temperature receptors. These factors may explain in part the improvement in correlation between sweat rate and temperature when skin and core temperatures are combined as mean body temperature. Hellon (5) has been able to find individual neurons in the anterior hypothalamus of the cat that respond to an increase of the temperature of either the neuron or the ambient air by an increase in frequency of impulses. Some neurons exhibited a distinct threshold of temperature below which the frequency was unchanged.

Foster and Weiner (6) have studied a subsystem in the cat consisting of the sweat glands of the footpad of the hind leg and their motor nerve supply in the internal plantar nerve. Electric stimulation of the nerve leads to an output of sweat measured by analysis of the water vapor in the air passing through a capsule enclosing the footpad. Figure 1 illustrates the relationship between the frequency of stimulation and the output of sweat. The neuroglandular transmitter substance in this system has been identified as acetylcholine, and it is assumed to be the same in man. Newman *et al.* (7) have shown that in the baboon evaporative cooling in a warm environment occurs by sweating rather than by panting. Since sweating is strongly inhibited by scopolamine, the baboon resembles cat and man in having cholinergic nerve endings in the sweat glands. In the burro (8) and horse (9), the sweat glands are mainly adrenergic. In the cat preparation described above, atropine acts as a reversible antagonist to acetylcholine at the receptors on the sweat glands. Atropine given intravenously inhibits sweating by an equal amount at both 2 and 4 stimuli/sec, and the effect is roughly proportional to the dose. This subsystem exhibits a set point and gain typical of the complete system. Atropine shifts the set point to the right in the figure with little change in gain or slope.

Much the same picture is seen in man when the environmental temperature is the stimulus acting on the whole system. In a neutral thermal environment the human eccrine sweat glands have little activity, and the effect of atropine on the

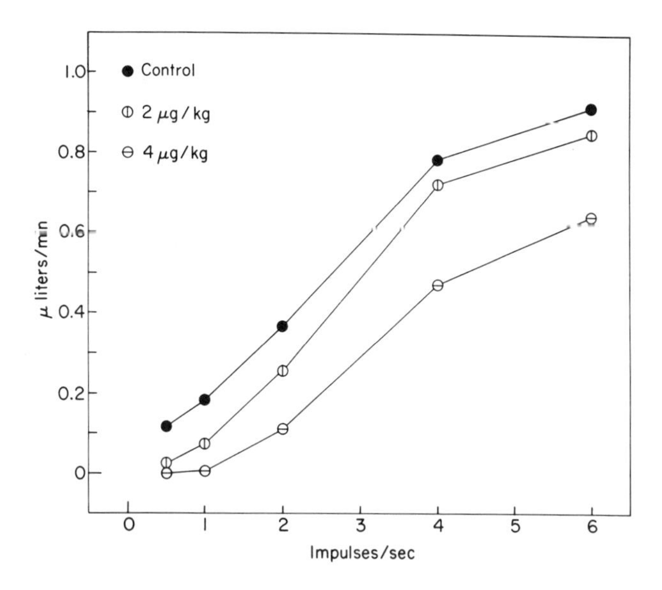

FIG. 1. Rate of sweating and rate of electrical stimulation with and without atropine sulfate. Recalculated data of Foster and Weiner (6).

body temperature is negligible (10); that is, atropine is not pyrogenic (Fig. 2). In a warm environment atropine produces a transient shutdown of sweating followed by an increase in body temperature and a recovery of sweating to the original rate (Fig. 3). The body temperature remains elevated until the atropine effect wears off (11). In Fig. 4 the sweat rate is plotted against mean body temperature; atropine shifts the line to the right (an increase in set point temperature) without much change in slope. But if atropine is acting only at the sweat glands, can this change in the set point be in the hypothalamus? Although a central effect of atropine is difficult to rule out, there is some indirect evidence. The blood-brain barrier is less permeable to the quaternary methyl salts of hyoscine and hyoscyamine than to the more common tertiary salts, of which atropine is an example. But the quaternary salts are stronger inhibitors of sweating than the tertiary salts when compared at the same molar dose (12). Thus, it is unlikely that the thermal effects of atropine are mediated centrally and atropine appears to provide us with a means of shifting the set point by peripheral action.

The results in the cat footpad preparation and in intact man suggest that a given dose of atropine antagonizes a given amount of transmitter and hence blocks the production of a corresponding amount of sweat. Let us apply this concept to the question of whether the sensitivity of the sweat glands to the transmitter changes with local temperature. When the environmental tempera-

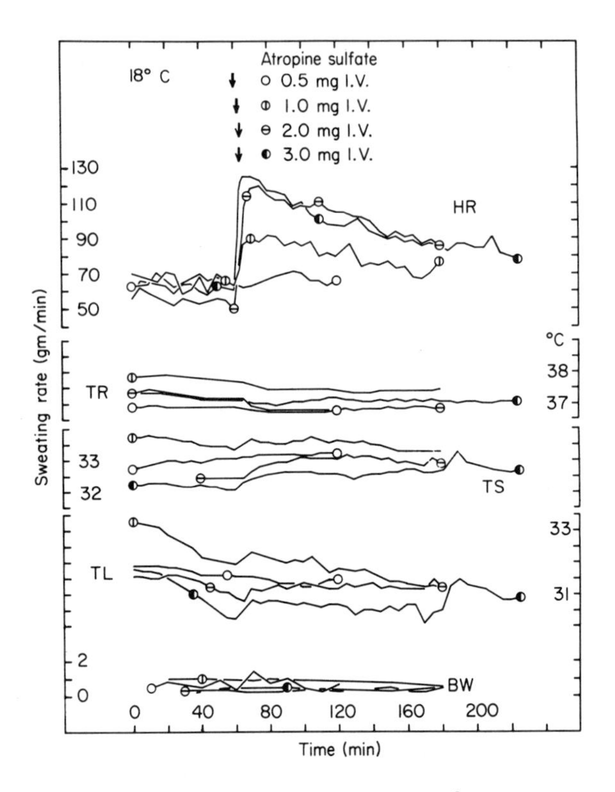

FIG. 2. Effects of atropine sulfate in one subject (FC) at 18°C. Arrows indicate times and doses of intravenous injection. In order from the top are heart rate (HR), rectal temperature (TR), mean skin temperature (TS), skin temperature of upper arm (TL), and body weight loss (BW).

ture is raised from 41° to 52°C the skin temperature may rise 1° or 2° and the sweat rate may double. How much of the increase in sweat rate is due to an increased release of transmitter substance and how much is due to an increase in the sensitivity of the glands to a given amount of transmitter? If the latter mechanism predominates, the inhibitory effect of an antagonist to the transmitter substance should be greater at the higher temperature. Atropine was infused slowly into a vein to provide a dose range over time. The inhibitory effect varied according to the dose. The experiment was performed on different days at 41° and 52°C. When the inhibition was plotted against dose (Fig. 5), the data for the two environments fell on the same line (13). Thus, according to this evidence, the increase in sweating with an increase in skin temperature was brought about mainly by an increase in the amount of transmitter rather than by a change in sensitivity.

The influence of the skin temperature upon the rate of sweat production has

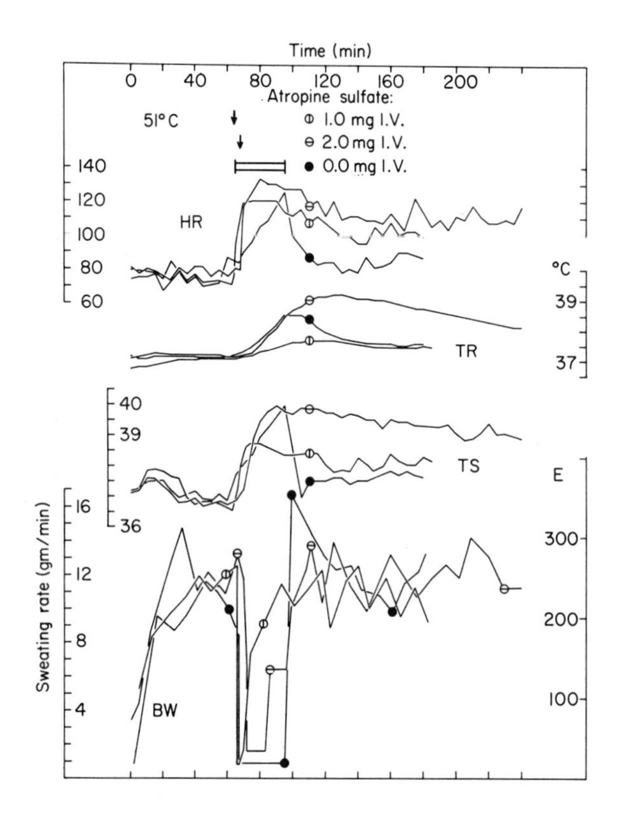

FIG. 3. Effect of atropine sulfate in one subject (FC) at 51°C. Arrows indicate times and doses of intravenous injection. In order from the top are heart rate (HR), rectal temperature (RT), mean skin temperature (TS), evaporative heat loss in calories per square meter per hour (E), and body weight loss (BW). The solid symbols refer to an experiment without atropine in which the subject was covered to reduce evaporation during the time indicated by the double bar.

been examined from other viewpoints. Local differences in recruitment of sweat glands have been described (14). Evidently there are local areas whose sweat glands and nervous connections each form a subsystem with its own set point and slope. In discussing these results, however, Wurster and McCook (15) have again called attention to the possibility that the sweating is affected by the rate of change of skin temperature as well as by the increment in temperature above the set point. There are three other ways in which local skin temperature may influence the local sweat rate (16). The first is through peripheral temperature receptors and the central nervous system. The second is through direct activation of the glands independent of neural stimulation; this may occur at temperatures above 42°C. The third is through a change in the amount of transmitter

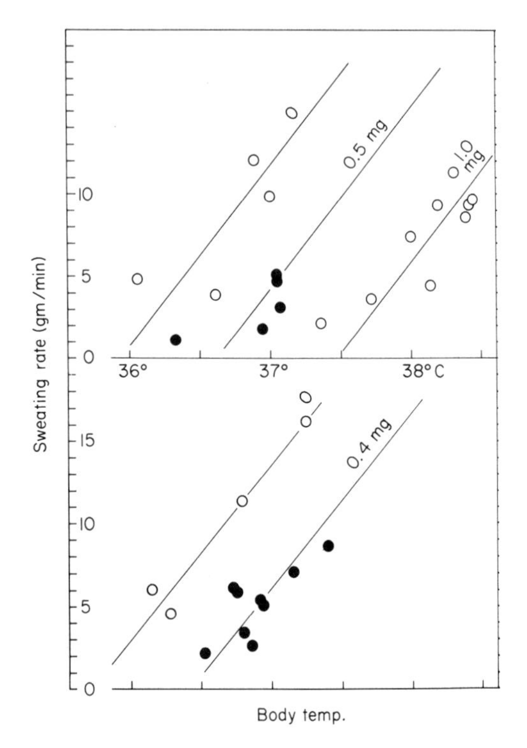

FIG. 4. Rate of sweating (gm/min) and mean body temperature in subjects 1 and 3 during brief weighing periods before and during slow intravenous infusions of atropine sulfate.

substance released by the nerve ending at the sweat gland for each nerve impulse, with the result that the higher the local temperature the more sweat is produced per nerve impulse. The most direct evidence for this third mechanism comes from records of sweating at two sites on the skin. Parallel cyclical variations in output are evidence of synchronous arrival of nerve impulses at the glands in the two sites. When one site is heated, the synchrony continues, but there is an increment in the amplitude of the cycles. However, the slope of the relationship between local sweat rate and local temperature has not been determined.

With this account of some of the elements of the control mechanism, we may proceed to examine a number of variables. Fever is perhaps the first variable in the study of temperature regulation. In his review, von Euler (2) noted that the concept of a change in set point in fever arose about a century ago. The site of the action of a pyrogen in evoking fever may be either central or peripheral. In their model, Stolwijk and Hardy (17) found it necessary to assume elevations in the set point both in the head core temperature and in the mean skin temperature. The effect of fever on the sweat rate-body temperature line has not

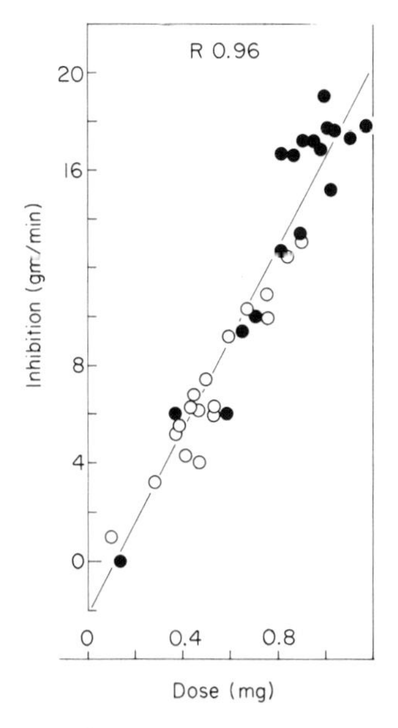

FIG. 5. Inhibition of sweating (gm/min) for various doses of atropine sulfate in one subject at environmental temperatures of 41° (○) and 52° C (●).

been defined. According to a recent abstract, the sensitivity of the thermoregulatory system is undiminished during steady-state fever (18).

In fever, for a given sweat rate the body temperature is higher than in health. In exercise, on the other hand, for a given sweat rate the body temperature is lower than at rest. Robinson (19) has recorded the rate of sweating at rest and in work in a number of thermal environments that provided a range of skin and rectal temperatures. When the rate was plotted against the mean body temperature, the curves for rest and work were parallel over most of the range; the intercept of the work curve on the temperature axis was shifted down by about 0.5°C. This indicates a shift in the set point somewhere in the system. It may be too soon to say whether exercise changes the slope of the sweat rate-body temperature relation (20). Gisolfi and Robinson (21) have given additional evidence of a neuromuscular factor in the regulation of sweating during exercise. Something of this sort is needed to account for the change in set point. The origin of this second variable remains obscure, although there is always the possibility of irradiation of impulses from the motor tracts exciting the central controlling mechanism for sweating. If the analogy with respiratory transients at the onset of work is worth anything, I am inclined to doubt that reflexes from muscle or joint mechanoreceptors are called into play (22).

A third variable that should not be overlooked is the inhalation of carbon dioxide. This increases thermal sweat production, but the mechanism is unknown (23).

A fourth variable is fatigue of the sweat glands. In men working in humid heat, the sweat rate markedly slows with elapsed time at a constant or rising body temperature (24, 25). Although the mechanism is not too clear, the decline with time in the output of the glands depends upon the wetness of the skin (26). Sweat rates and body temperatures have been recorded on a group of men at rest in still air at 41° and 52°C. The lines joining the data for the two environmental temperatures show a significant decline in slope with each successive hour (Fig. 6). I presented the data in this way to illustrate a change in the slope of the sweat rate-body temperature line that probably is peripheral in origin (27).

The state of hydration of the body is a fifth variable. Space does not permit a review of the extensive literature dealing with the complex interrelationships between water and salt balance and the composition and rate of secretion of

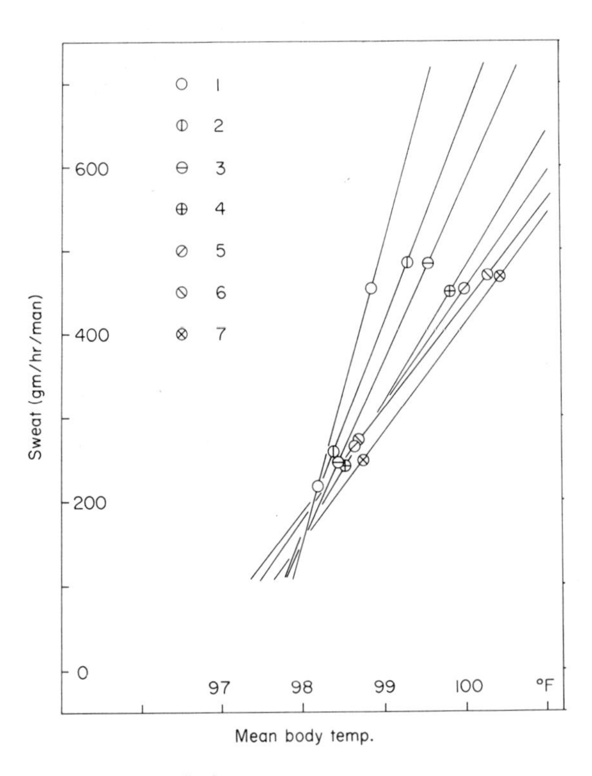

FIG. 6. Sweat production (gm/hr/man) and mean body temperature. Environmental conditions with minimal air movement were 41°C dry bulb and 21°C wet bulb for the lower group of points and 52°C dry bulb and 26°C wet bulb for the upper points. The numbered symbols identify means in 7 successive hours for 14 men at rest.

sweat. Evidence from Moroff and Bass (28) indicates that overhydration slightly increases the sweat rate and lowers the body temperature; Pearcy *et al.* (29) show that dehydration can reduce the sweat rate by 15% and increase the body temperatures. I did not find data for a range of body temperatures that would permit definition of the relation between sweat rate and body temperature for different states of hydration.

Acclimatization to heat has continued to be a popular subject for investigation. The critical factor is an elevation of body temperature repeated many times. This may be achieved passively in either a warm bath or a heated suit according to the procedure at Mill Hill (30) or by hard work in a neutral thermal environment (31). The relationship between sweat rate and body temperature in different samples of gold miners before and after acclimatization has been described by Wyndham (32). His data show a shift in the set point to a lower temperature and a rise in the maximum value of the sweat rate. In another arrangement of the same data there is an increase in the slope with acclimatization. This would indicate that at some point in the system a given increase in body temperature is more effective in stimulating the glands after acclimatization than before. By means of the inhibitory effect of atropine one link in the system can be tested–the interaction between the transmitter substance and the sweat gland. If in the course of repeated exposures to heat the sweat glands themselves adapt to increase their output in response to each unit of transmitter, as Fox *et al.* (33) suggest, then the inhibitory effect of a given dose of atropine should increase after acclimatization. If the sensitivity of the sweat glands is unchanged but a given increase in body temperature results in a greater release of transmitter per nerve impulse, or in a greater number of nerve impulses, then the proportionality between transmitter and sweat gland activity is maintained and the inhibitory effect of atropine should be unchanged.

In four volunteers, the body temperatures and rates of sweating were measured before, during, and after a slow intravenous infusion of atropine sulfate (27). Two tests were made on each man, one at 41°, the other at 52°C. During the next 2 weeks the men were exposed to 52°C for 2 hr on 8 days with some treadmill exercise. In the fourth week the atropine tests were repeated at 41° and 52°C. The degree of acclimatization achieved is shown in Fig. 7. Although the sweat rates did not change greatly, in each man the skin and rectal temperatures were reduced in both environmental conditions. The inhibitory effects are presented in Fig. 8 as regressions on dose. The doses were well scattered over the range from 0 to 1 mg/man. The regression slope and hence the inhibitory effect of atropine was not significantly altered by acclimatization under these conditions. Thus, the response of the sweat glands to a unit amount of neuroglandular transmitter substance was unaffected by acclimatization. This is consistent with the observations in the absence of atropine during which the slopes of the lines relating sweat rate with either skin or rectal temperature were

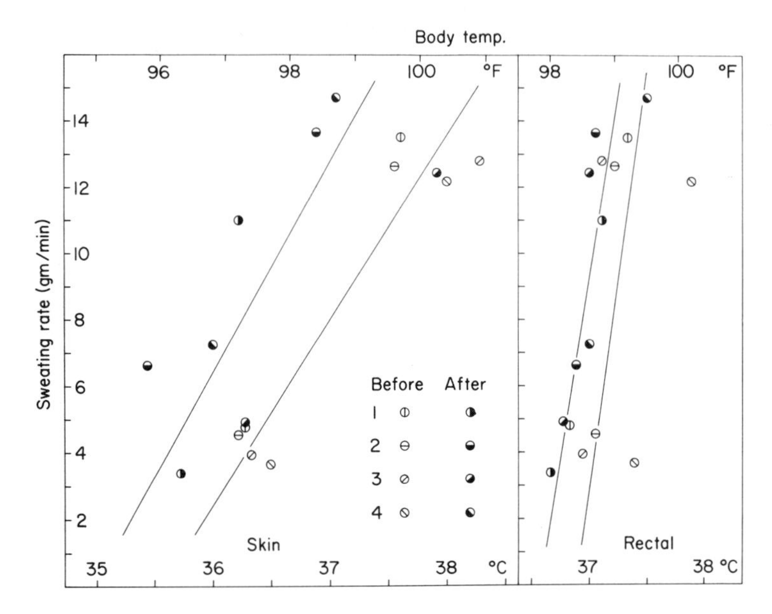

FIG. 7. Sweat rate, mean skin temperature, and rectal temperature in four subjects at 41° and 52°C before and after acclimatization to heat.

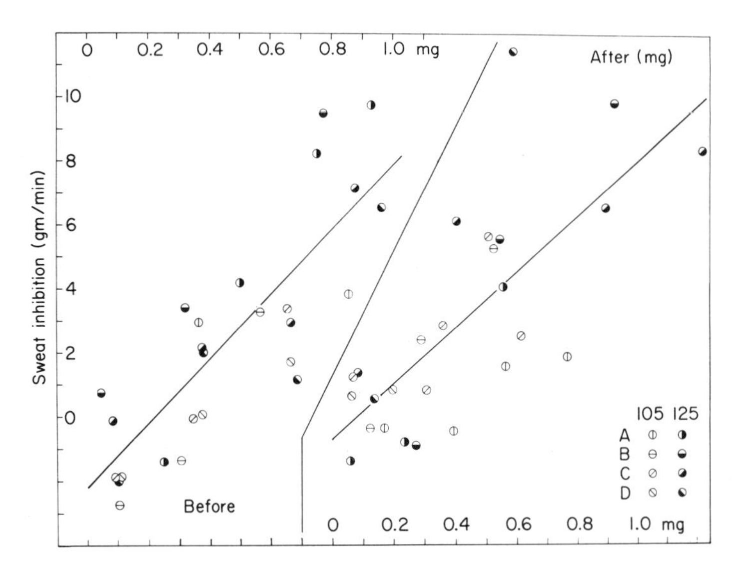

FIG. 8. Inhibition of sweating and dose of atropine sulfate before and after acclimatization to heat. Each point represents a brief weighing period during slow intravenous infusions at 41°C (105°F) and 52°C (125°F) in the four subjects shown in Fig. 7.

unchanged. Extrapolation of the lines back to zero sweat rate would indicate a change in set point. However, there is no indication of which subsystem is responsible for the change.

References

1. Wolfe, S., Cage, G., Epstein, M., Tice, L., Miller, H., and Gordon, R. S., Jr. (1970). Metabolic studies of isolated human eccrine sweat glands. *J. Clin. Invest.* **49**, 1880.
2. von Euler, C. (1961). Physiology and pharmacology of temperature regulation. *Pharmacol. Rev.* **13**, 361.
3. Hardy, J. D. (1961). Physiology of temperature regulation. *Physiol. Rev.* **41**, 521.
4. Hensel, H. (1971). Temperature receptors in the skin. *In* "Physiological and Behavioral Temperature Regulation" (J. D. Hardy *et al.*, eds.), Chapter 30, p. 442. Thomas, Springfield, Illinois.
5. Hellon, R. F. (1970). Hypothalamic neurones responding to changes in hypothalamic and ambient temperatures. *In* "Physiological and Behavioral Temperature Regulation" (J. D. Hardy *et al.*, eds.), Chapter 32, p. 463. Thomas, Springfield, Illinois.
6. Foster, K. G., and Weiner, J. S. (1970). Effects of cholinergic and adrenergic blocking agents on the activity of the eccrine sweat glands. *J. Physiol. (London)* **210**, 883.
7. Newman, L. M., Cummings, E. G., Miller, J. L., and Wright, H. (1970). Thermoregulatory responses of baboons exposed to heat stress and scopolamine. *Physiologist* **13**, 271.
8. Bullard, R. W., Dill, D. B., and Yousef, M. K. (1970). Responses of the burro to desert heat stress. *J. Appl. Physiol.* **29**, 159.
9. Bell, F. R., and Evans, C. L. (1956). The relation between sweating and the innervation of sweat glands in the horse. *J. Physiol. (London)* **134**, 421.
10. Craig, F. N. (1952). Effects of atropine, work and heat on heart rate and sweat production in man. *J. Appl. Physiol.* **4**, 826.
11. Craig, F. N., and Cummings, E. G. (1965). Speed of action of atropine on sweating. *J. Appl. Physiol.* **20**, 311.
12. Craig, F. N. (1970). Inhibition of sweating by salts of hyoscine and hyoscyamine. *J. Appl. Physiol.* **28**, 779.
13. Cummings, E. G., and Craig, F. N. (1967). Influence of the rate of sweating on the inhibitory dose of atropine. *J. Appl. Physiol.* **22**, 648.
14. Wurster, R. D., Hassler, C. R., McCook, R. D., and Randall, W. C. (1969). Reversal in patterns of sweat recruitment. *J. Appl. Physiol.* **26**, 89.
15. Wurster, R. D., and McCook, R. D. (1969). Influence of rate of change in skin temperature on sweating. *J. Appl. Physiol.* **27**, 237.
16. MacIntyre, B. A., Bullard, R. W., Banerjee, M., and Elizondo, R. (1968). Mechanism of enhancement of eccrine sweating by localized heating. *J. Appl. Physiol.* **25**, 255.
17. Stolwijk, J. A. J., and Hardy, J. D. (1966). Temperature regulation in man–a theoretical study. *Pfluegers Arch. Gesamte Physiol. Menschen Tiere* **291**, 129.
18. Cunningham, D. J., and Stolwijk, J. A. J. (1971). A comparison of thermoregulatory responses in the afebrile and febrile state. *Fed. Proc., Fed. Amer. Soc. Exp. Biol.* **30**, 319.
19. Robinson, S. (1963). Temperature regulation in exercise. *Pediatrics* **32**, 691.
20. Saltin, B., Gagge, A. P., and Stolwijk, J. A. J. (1970). Body temperatures and sweating during thermal transients caused by exercise. *J. Appl. Physiol.* **28**, 318.

21. Gisolfi, C., and Robinson, S. (1970). Central and peripheral stimuli regulating sweating during intermittent work in men. *J. Appl. Physiol.* **29**, 761.
22. Craig, F. N., Cummings, E. G., and Blevins, W. V. (1963). Regulation of breathing at beginning of exercise. *J. Appl. Physiol.* **18**, 1183.
23. Bullard, R. W. (1964). Effects of carbon dioxide inhalation on sweating. *J. Appl. Physiol.* **19**, 137.
24. Robinson, S. (1949). Physiological adjustments to heat. *In* "Physiology of Heat Regulation and the Science of Clothing" (L. H. Newburgh, ed.), Chapter 5, p. 193. Saunders, Philadelphia, Pennsylvania.
25. Wyndham, C. H., Strydom, N. B., Morrison, J. F., Williams, C. B., Bredill, G. A. G., and Peter, J. (1966). Fatigue of the sweat gland response. *J. Appl. Physiol.* **21**, 107.
26. Hertig, B. A., Riedesel, M. L., and Belding, H. S. (1961). Sweating in hot baths. *J. Appl. Physiol.* **16**, 647.
27. Craig, F. N., Cummings, E. G., Froehlich, H. L., and Robinson, P. F. (1969). Inhibition of sweating by atropine before and after acclimatization to heat. *J. Appl. Physiol.* **27**, 498.
28. Moroff, S. V., and Bass, D. E. (1965). Effects of overhydration on man's physiological responses to work in heat. *J. Appl. Physiol.* **20**, 267.
29. Pearcy, M. S., Robinson, S., Miller, D. I., Thomas, J. T., Jr., and De Brota, J. (1956). Effects of dehydration, salt depletion and pitressin on sweat rate and urine flow. *J. Appl. Physiol.* **8**, 621.
30. Fox, R. H., Goldsmith, R., Hampton, I. F. G., and Lewis, H. E. (1964). The nature of the increase in sweating capacity produced by heat acclimatization. *J. Physiol. (London)* **171**, 368.
31. Piwonka, R. W., and Robinson, S. (1967). Acclimatization of highly trained men to work in severe heat. *J. Appl. Physiol.* **22**, 9.
32. Wyndham, C. H. (1967). Effect of acclimatization of the sweat rate/rectal temperature relationship. *J. Appl. Physiol.* **22**, 27.
33. Fox, R. H., Goldsmith, R., Kidd, D. J., and Lewis, H. E. (1963). Acclimatization to heat in man by controlled elevation of body temperature. *J. Physiol. (London)* **166**, 530.

V

Some Comments on Low Temperatures in the Desert and at Altitudes

L. D. CARLSON

A discussion of responses to cold in desert animals seems trivial and somewhat apart from the main theme of this book and the main environmental stress to which desert animals are exposed. However, a casual look at most travel advertisements for the desert will impress one with the emphasis on the large changes in temeprature from day to night. In addition, many desert animals live in the mountains nearby as well, and the altitude calls for adaptive responses both to pressure and the daily temperature variations (Fig. 1) (1).

Temperature regulation in desert animals has been reviewed (1a-3) as has the effect of altitude on animals (2). The intent of this chapter is to discuss some of the physiological systems that are periodically challenged by temperature and that are receptive to the environmental stimuli in an adaptive way. The temporal and intensity effects of these stresses on animals are the focus of interest.

My analysis is speculative in some aspects and may be over generalized, but what I hope for is that it will carry a small fraction of the stimulus to investigation that Dr. Dill provided in his book, "Life, Heat and High Altitude" (4), and perhaps, in the spirit of this book, it will open some questions for investigation.

In his book, "Physiological Regulations," Adolph (5) introduced the concept of "load" in physiological regulations and presented 13 theories, some of which are relevant to the following discussion. I prefer to give these the rank of principles rather than theories.

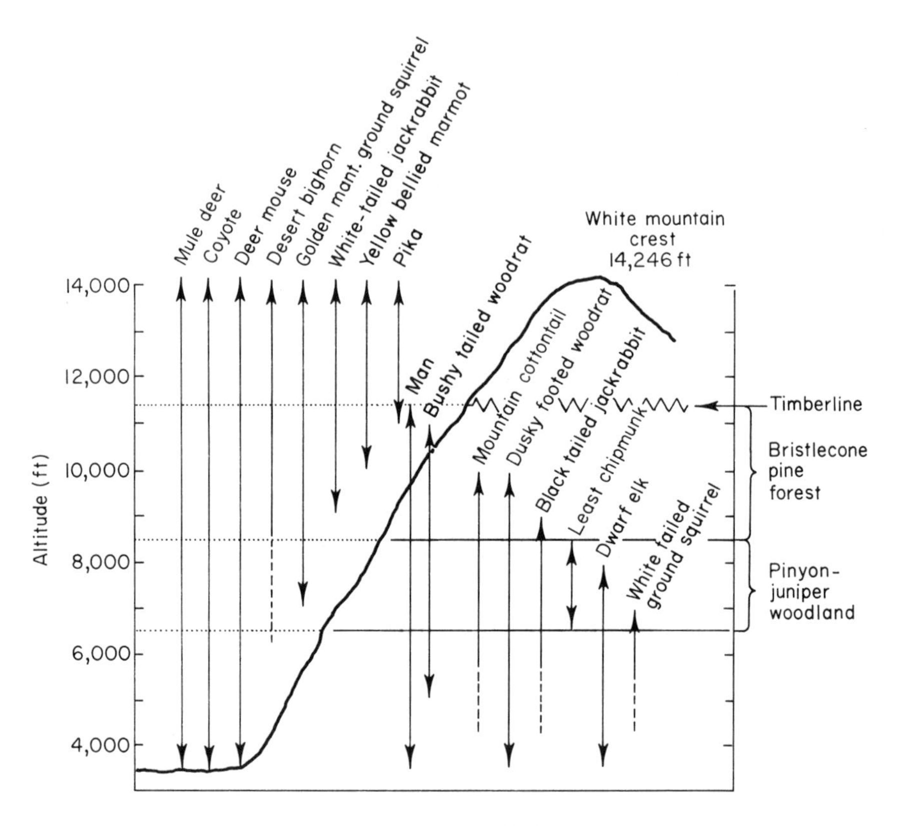

FIG. 1. An indication of the range of altitudes in which man and animals survive (1).

PRINCIPLE OF PHYSIOLOGICAL PURPOSE

All vital mechanisms, varied as they are, have only one object, that of preserving the internal environment (6). However, Bernard's principle should not be thought of as absolute constancy, and it should be understood that variations in the properties of the internal environment may be both cyclical and adaptive (7).

PRINCIPLE OF NECESSARY REGULATIONS

All living units regulate or maintain certain properties within limits. These regulations have the property of an equilibration–balance between displacement (strain) and load (stress).

PRINCIPLE OF INTERDEPENDENCY OF REGULATIONS

No component is regulated independently of all others.

PRINCIPLE OF PHYSIOLOGICAL MAGNITUDES

Loads and exchanges must be viewed with respect to appropriate time constants for the systems involved.

PRINCIPLE OF SCALING

Response times and load magnitude bear some relation to body size and metabolic turnover rates.

PRINCIPLE OF HYSTERETIC EFFECT

The previous history or treatment of a body influences the subsequent response to a given force or changed condition. A biological system is inconceivable without its external environment.

There are broad generalizations about climatic factors in evolution reviewed by Baker (8), by Newman (9), and by Barnicott (10), but this discussion will concern systems responses that may be time-dependent within the time range of day-to-day existence. In passing, it may be worthy of note that perhaps the vagueness of inferences that can be drawn about climatic factors in evolution is due to the usual emphasis on the average climatic condition, as contrasted to deductions that might be inferred from a range or time-intensity analysis.

As a frame of reference for discussion of the effects of low temperature and/or low pressure on animals who also are exposed to heat, the interaction of some elements of the external system with certain components of the internal system are illustrated in Fig. 2. The external environment is characterized by its temperature and partial pressure of oxygen. In the case of temperature, rate of change is also indicated as an effector. These environmental factors change conditions at the skin as far as temperature is concerned and influence internal temperature. The combined changes reflect, via the hypothalamus, to initiate activation of the adrenal medulla and changes in activity of the autonomic nervous system. These same changes may reflect, via the hypothalamus, to provide changes in releasing factors to stimulate ACTH secretion or TSH secretion with resulting changes in blood concentration. The effectors in

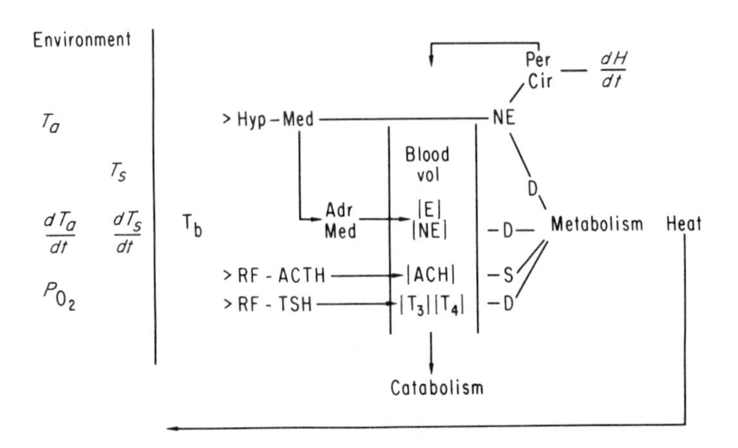

FIG. 2. A schematic to represent the interaction of environment with certain physiological systems. T_a is the air temperature; t, time; P_{O_2}, partial pressure of oxygen; T_s, skin temperature; T_b, body temperature; Hyp, hypothalamus; Med, medulla; RF, releasing factor; ACTH, adrenocorticotrophic hormone; TSH, thyroid stimulating hormone; Adr Med, adrenal medulla; Per Circ, peripheral circulation; E, epinephrine; NE, norepinephrine; H, heat; ACH, adrenocortical hormones; T_3, tricodothyronine; T_4, thyroxine.

temperature regulation are the peripheral circulation (conductance) and metabolism (heat production).

In order to direct attention to some of the questions relevant to considerations of cold exposure, I will establish a series of paradigms with a brief discussion of each. This is not an inclusive list, but it is intended to indicate some areas of interest.

1. The ambient temperature at which a minimum resting metabolism will occur may be related to the temperature of existence. Brody (11) has presented data to support this possibility, although Scholander *et al.* (12), in their classic paper, do not discuss it. The acclimation procedure (exposure to a fixed environmental temperature) is a different stimulus than the natural environment, and the desert or mountain existence provides a uniquely wide variation in daily temperatures for study of the mammal in which the daily patterns are integrated into the systems response. The critical criteria for study may be derived from Brody's diagrammatic presentation (11) (Fig. 3), and the physiological evaluation must take cognizance of factors outlined by Wyndham *et al.* (13). That is to say, the experimental definition must include the determination of the critical air temperatures at which the main temperature regulatory reactions occur and establish the form of the functional relationships.

2. The response to an increased temperature or positive heat load is related to previous exposures to that temperature. Acclimatization to heat is a dramatic

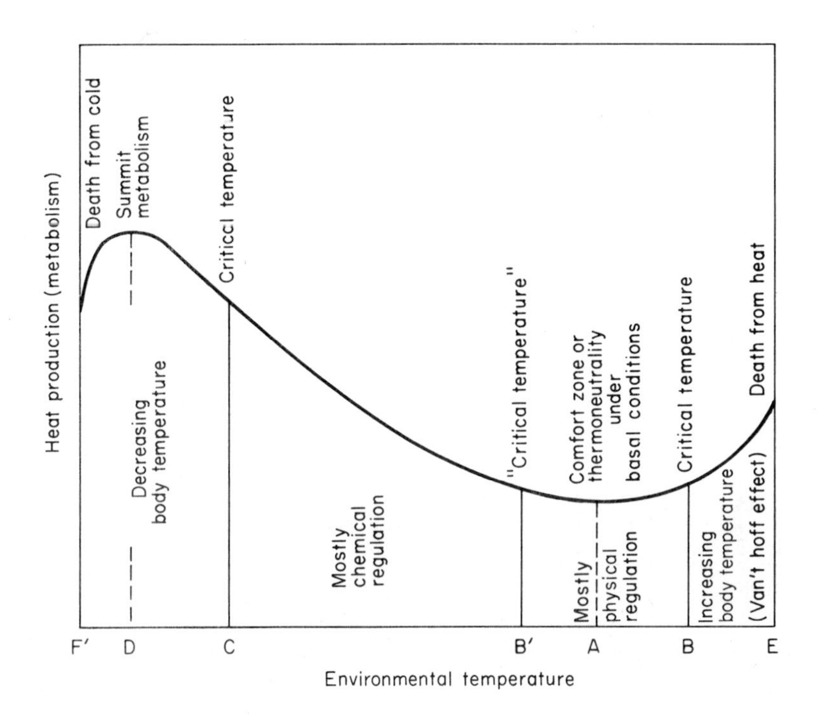

FIG. 3. A schematic representation of the metabolic and physical systems involved in temperature regulation (11).

example of adaptation to a changed environment. The acclimatization requires 4-7 days in man. The major parameters changed are the rate of sweating (increased) and the change in rectal temperature and heart rate (both decreased). During the acclimatization period, there are a number of transient responses. Cardiac output, blood volume, venous tone, and peripheral blood flow are parameters that show a transient increase (14). Fox *et al.* (15) brought about head acclimatization by short (2–4 hr) periods of raised rectal temperatures, while the subjects were spending the remainder of the day in the environment of the British Isles. Lind and Bass (16) reported that men who walked for 100 min continuously each day acclimated more effectively than men who walked for only 50 min each morning or those who walked each morning and again each afternoon for 50 min at a time (a total of 100 min each day). Once acquired, acclimatization to heat is retained for 2 weeks or more.

3. The response to low temperature or a negative heat load of a change in conductance (primarily due to peripheral blood flow alterations) is related to

previous temperature exposures. The now classic example (though it is still questioned as a generality) is the change in the response to low temperature in the hands of fish filleters or persons habitually exposed to cold (17, 18). Another example is the observation that blood flow in the rabbit ear has a different pattern of response to temperature when animals are cold acclimated (19), a change that includes an altered response to infused norepinephrine.

Similarly, the metabolic response to a cold exposure is dependent on previous cold exposure. The altered metabolic response to cold exposure is more than a quantitative change, since it involves a change in the stimulus-response pathways, that is to say, the shivering versus nonshivering response. It is intriguing that the nonshivering response is present in newborn animals—gradually supplanted by the shivering response as the animal matures. Exposure to cold brings the response back into effect. The time course of this action of cold exposure, as determined by the metabolic effect of norepinephrine, is a matter of weeks (Fig. 4) (19a); the magnitude is dependent on the temperature (Fig. 5) (19a); and the effect of cold exposure is evident with as little as a 6-hr-a-day exposure (Fig. 6) (19b). The nonshivering mechanism shows a specificity for norepinephrine that is not evident in *in vitro* experiments. The effects of norepinephrine and epinephrine are not distinguishable in their effect on cyclic AMP (Fig. 7) (19c), but the metabolic effect *in vivo* is markedly different (Fig. 4). Thus, not only the "switching" in the stimulus-response path, but also the mode of action of metabolic action (most probably in adipose tissue) remain to be elucidated (20).

4. Although attractive in teleological arguments, the role of the thyroid in response to temperature is still unclear. The studies of Dempsey and Astwood (21) showed that the requisite amount of thyroxin to prevent hypertrophy in the presence of a thiouracil block varies with temperature, the requisite amount of hormone in rats being 1.7 μg/day at 35°C, 5.2 μg at 25°C, and 9.2 μg at 1°C, and other studies have subsequently confirmed this finding (22). Hsieh and Carlson (23) showed that, in the absence of thyroid hormone, the animal cannot achieve the metabolic rates necessary to maintain thermoregulation. The synergistic effect of epinephrine and thyroxin has further been shown (24) and again presents evidence for an effective thyroxin level that is temperature dependent. The direct relationship of thyroid to temperature regulation, however, has been questioned (3).

5. Corticosteroids have a supporting but transient role in reactions to environmental stimuli. The characterization of the response of the adrenocortical gland by Sayers and Sayers (25) still seems valid (Fig. 8) (25a, b). The response to corticosteroid injection was used as a means of characterizing the status of physiological systems by Hale and Mefford (26), who found the toxic effects of thyroxin and corticosteroid to be enhanced in hot environments and reduced in cold environments. In this area of research, it is particularly important to be aware of the circadian periodicity (27).

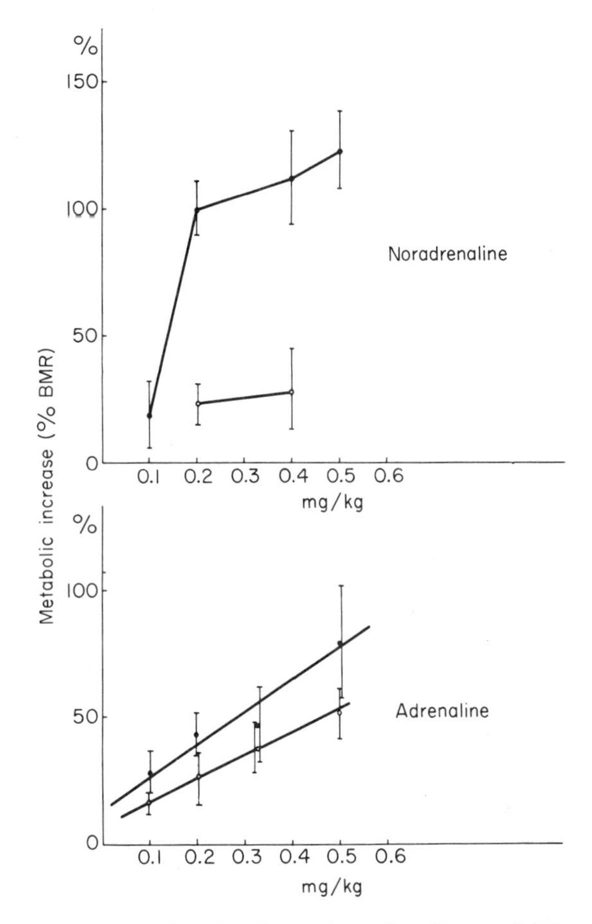

FIG. 4. Dose-response curves for adrenaline and noradrenaline as related to environmental temperatures. Upper, metabolic increase related to intramuscular injections of noradrenaline. Lower, metabolic increase related to intramuscular injections of adrenaline (19a). •, *CA(+5°C);* ○, *WA(+25°C).*

6. The cellular basis of the response to temperature has been studied at the enzyme level, with the result that acclimatization can be characterized by enzyme changes (28). Most significant in the cellular studies has been the elucidation of the role of adipose tissue, especially brown adipose tissue (29).

What is the application of these principles as exemplified by the paradigms selected? The general view, that the organism's reaction to the environment in time is important in order to sum various stimuli with appropriate time constants, may be too simplistic, but it directs attention to the tenuous nature of some of the accepted views that are derived from short exposures to the

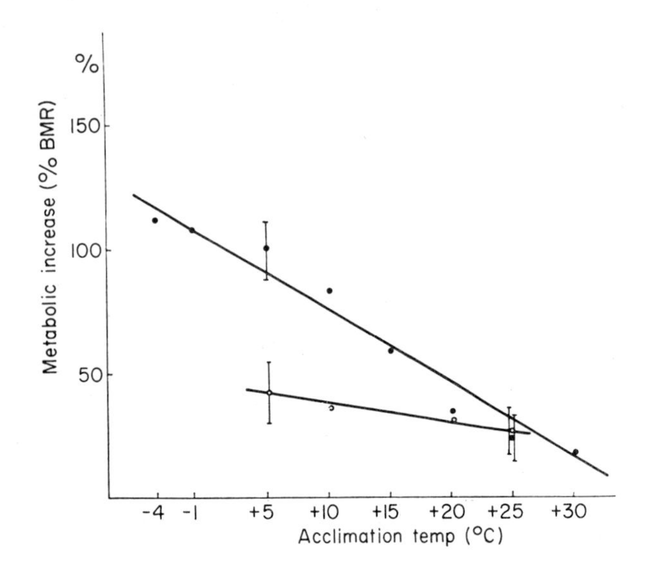

FIG. 5. Effect of acclimation temperature on the metabolic response after intramuscular injections of noradrenaline and adrenaline in anesthetized rats, kept at each temperature for 3 weeks prior to experiments (19a). •, 0.2 mg NA/kg I.M.; ○, 0.2 mg A/kg I.M.

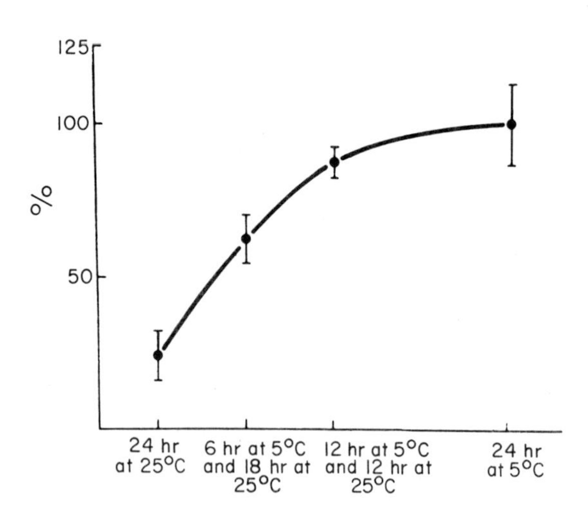

FIG. 6. The partitioning of the effect of cold exposure at a dose of 0.2 mg/kg of noradrenaline (19b).

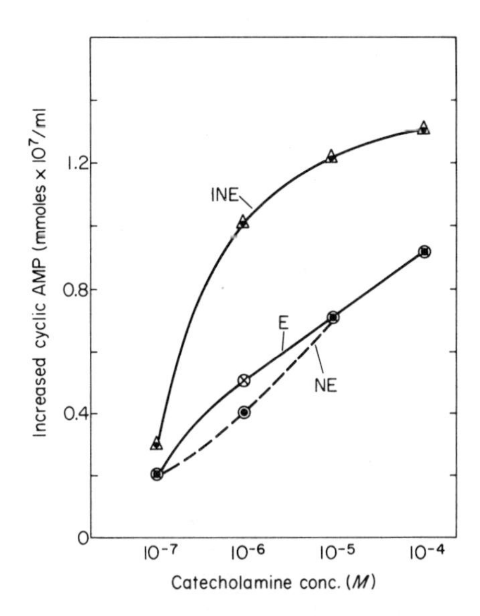

FIG. 7. The effect of various concentrations of the catecholamines on cyclic AMP accumulation by a homogenate of rat epididymal fat. INE, isoproterenol; E, epinephrine; and NE, norepinephrine (19c).

environment or the study of the obvious or dominant environmental extreme. There may be a hierarchy in environmental effects; e.g., heat seems to have dominance over cold exposures if intermittent.

Recently, Brück *et al.*, (30) have characterized the acclimatization phenomenon in terms of the set point for temperature regulation. The set point of the temperature control system in guinea pigs is at a lower level in cold acclimatized animals. In animals exposed to 12-hr warm and 12-hr cold, there was a widening of the inter-threshold zone; the shivering threshold was decreased but the heat polypnea threshold was not changed.

Animals in the desert and at altitude must be considered to be exposed to cycling temperatures, and the partitioning effect on the regulatory systems will give evidence as to which responses are dominant. The idea that the regulation zone may widen as a result of these cycles opens an interesting area for exploration.

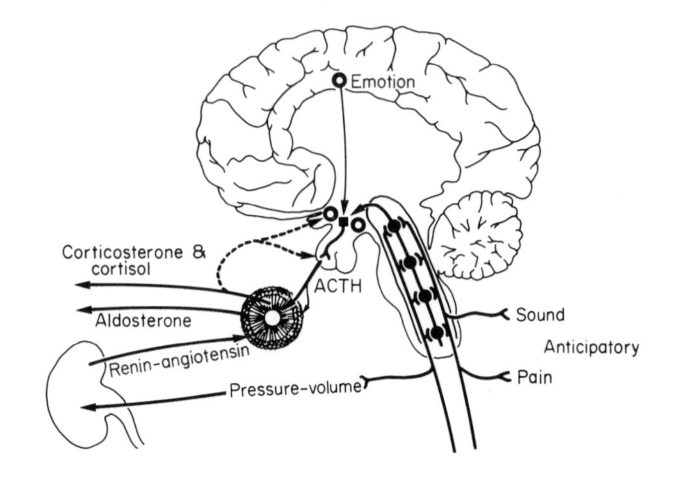

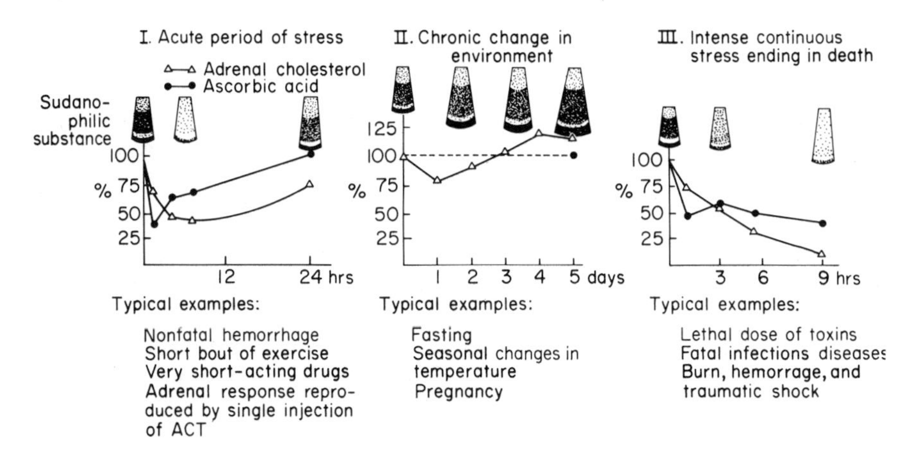

FIG. 8. A schematic representation of the neuroendocrine relationship to activity of the adrenal cortex. [(The neural relation from Sayers and Sayers (25a) and the adrenal cortical gland adapted from Sayers and Travis (25b).] Three types of responses are shown with examples.

References

1. Folk, G. E. (1966). "Introduction to Environmental Physiology." Lea and Febiger, Philadelphia, Pennsylvania.

1a. Bartholomew, G. A., and Dawson, W. R. (1968). Temperature regulation in desert mammals. *In* "Desert Biology" (G. W. Brown, ed.), Vol. 1, Chapter VIII, p. 396-424. Academic Press, New York.

2. Dill, D. B., ed. (1964). "Handbook of Physiology." Sect. 4. Williams & Wilkins, Baltimore, Maryland.

3. Chaffee, R. R. J., and Roberts, J. C. (1971). Temperature acclimation in birds and animals. *Annu. Rev. Physiol.* **33**, 155.

4. Dill, D. B. (1938). "Life, Heat and Altitude." Harvard Univ. Press, Cambridge, Massachusetts.

5. Adolph, E. F. (1943). "Physiological Regulations." Cattell, Lancaster, Pennsylvania.

6. Bernard, C. (1878). "Leçons sur des phénomènes de la vie communs aux animaux et aux végétaux." Baillière et Fils, Paris.

7. Bernard, C. (1878). "An Introduction to Experimental Medicine" Henry Schuman, New York. (transl. by H. C. Greene, 1949).

8. Baker, P. T. (1960). Climate, culture and evolution. *Human Biol.* **32**, 3.

9. Newman, N. T. (1961). Biological adaptations of man to his environment: Heat, cold, altitude and nutrition. *Ann. N. Y. Acad. Sci.* **91**, 617.

10. Barnicott, N. A. (1959). Climatic factors in the evolution of human populations. *Cold Spring Harbor Symp. Quant. Biol.* **24**, 115.

11. Brody, S. (1945). "Bioenergetics and Growth." Reinhold, New York.

12. Scholander, P. F., Hock, R., Walters, V., Johnson, F., and Irving, L. (1950). Heat regulation in some Arctic and tropical birds and mammals. *Biol. Bull.* **99**, 237.

13. Wyndham, C. H., Carlson, L. D., Morrison, J. F., Krog, J., and Eagen, C. (1966). A comparison of two test procedures of cold responses. *In* "Human Adaptability and Its Methodology" (H. Yoshimura and J. S. Weiner, eds.), pp. 18-35, Japan Society for the Promotion of Sciences, Tokyo.

14. Leithead, C. S., and Lind, A. R. (1964). "Heat Stress and Heat Disorders." Davis, Philadelphia, Pennsylvania.

15. Fox, R. H., Goldsmith, R., Kidd, D. J., and Lewis, H. E. (1963). Acclimatization to heat in man by controlled elevation of body temperature. *J. Physiol. (London)* **166**, 530.

16. Lind, A. R., and Bass, D. E. (1963). The optimal exposure time for development of acclimatization to heat. *Fed. Proc., Fed. Amer. Soc. Exp. Biol.* **22**, 704.

17. Nelms, J. D., and Soper, J. G. (1962). Cold vasodilatation and cold acclimatization in the hands of British fish filleters. *J. Appl. Physiol.* **17**, 444.

18. Krog, J., Folkow, B., Fox, R. H., and Andersen, K. L. (1960). Hand circulation in the cold of Lapps and North Norwegian fishermen. *J. Appl. Physiol.* **15**, 654.

19. Nagasaka, T., and Carlson, L. D. (1971). Effects of blood temperature and perfused norepinephrine on vascular responses of the rabbit ear. *Amer. J. Physiol.* **220**, 289.

19a. Jansky, L. (1969). a Interspecies Differences in cold adaptation in shivering and non-shivering thermogenesis. *Fed. Proc., Fed. Amer. Soc. Exp. Biol.* **28**, 1053.

19b. Jansky, L., Bartunkova, R., and Zersberger, E. (1967). Acclimation of the white rat to cold; Noradrenaline thermogenesis. *Physiol. Bohemoslov.* **16**, 366.

19c. Butcher, R. W., and Sutherland, E. A. (1967). The effects of the catecholamines, adrenergic blocking agents, prostoglandin E, and insulin on cyclic AMP levels in the rat epidiymal fat pad in vitro. *Ann. N. Y. Acad. Sci.* **139**, 849.

20. Himms-Hagen, J. (1970). Adrenergic receptors for metabolic response in adipose tissue. *Fed. Proc., Fed. Amer. Soc. Exp. Biol.* **29**, 1388.
21. Dempsey, E. W., and Astwood, E. B. (1943). Determination of the rate of thyroid hormone secretion at various environmental temperatures. *Endocrinology* **32**, 509.
22. Carlson, L. D. (1960). Nonshivering thermogenesis and its endocrine control. *Fed. Proc., Fed. Amer. Soc. Exp. Biol.* **19**, 25.
23. Hsieh, A. C. L., and Carlson, L. D. (1957). Role of the thyroid in metabolic response to low temperature. *Amer. J. Physiol.* **188**, 40.
24. Swanson, H. (1957). The effect of temperature on the potentiation of adrenalin by thyroxine in the albino rat. *Endocrinology* **60**, 205.

25a. Sayers, G., and Sayers, M. (1949). The pituitary-adrenal system. *Ann. N. Y. Acad. Sci.* **50**, 522.

25b. Sayers, G., and Travis, R. H. (1970). The Pituitary-adrenal system. *In* "The Pharmacological Basis of Therapeutics" (L. S. Goodman and A. Gilman, eds.) 4th ed., Chapter 72, p. 1604-1642. Macmillan, New York.

26. Hale, H. B., and Mefford, R. B., Jr. (1959). Effects of adrenocorticotropin on temperature- and pressure-dependent metabolic functions. *Amer. J. Physiol.* **197**, 1291.
27. Halberg, F. (1969). Chronobiology. *Annu. Rev. Physiol.* **31**, 675.
28. Masoro, E. J. (1966). Effect of cold on metabolic use of lipids. *Physiol. Rev.* **46**, 67.
29. Smith, R. E., and Horwitz, B. A. (1969). Brown fat and thermogenesis. *Physiol. Rev.* **49**, 330.
30. Brück, K., Winnenberg, W., Gallmeier, H., and Zehm, B. (1970). Shift of threshold for shivering and heat polypnea as a mode of thermal adaptation, *Pfluegers arch.* **321**, 159.

VI

Cardiovascular and Respiratory Reactions to Heat

S. ROBINSON

Introduction

The strain shown by a man exposed to hot environments may be indicated by increments in his heart rate, sweating, skin temperature, and rectal temperature. Circulatory strain is indicated by increased heart rate and, under conditions nearing the limits of tolerance, by a fall in blood pressure and syncope if the subject attempts to stand. The effects of increasing heat stress on men are shown in Fig. 1, representing data from experiments in which two men who were well acclimatized to work in the heat walked on the treadmill in environments where the temperature and relative humidity were varied (1). At the lowest values of relative humidity used in the study, the men's skin temperatures and rates of sweating in these 2-hr walks were both elevated significantly with each 5°C increase in dry bulb temperature between 23° and 50°C; in the same experiments neither heart rate nor rectal temperature was elevated to any important degree by increasing the dry bulb temperature up to 50°C with a relative humidity of 15%. At a dry bulb temperature of 23°C, raising the humidity had no effect on any of the functions. At 28°C the men's rates of sweating and skin temperatures increased curvilinearly with increasing humidity, but neither heart rate nor rectal temperature was increased significantly above their values at 23°C. The effects of increasing relative humidity on all four of the processes became more and more pronounced as the dry bulb temperature was raised above 28°C. Significant circulatory strain in the working men, manifested

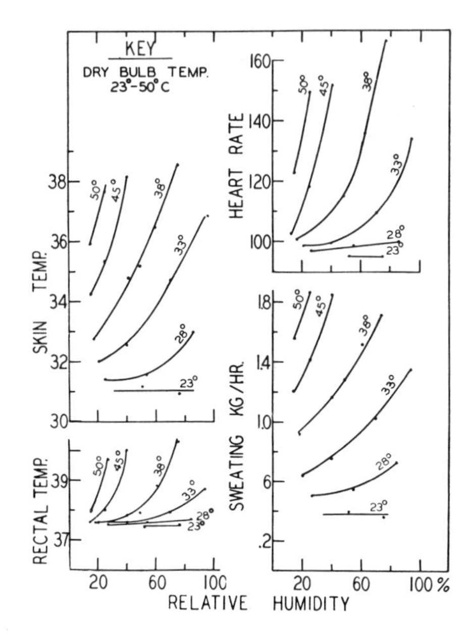

FIG. 1. Effects of varying relative humidity at six different dry bulb temperatures on men performing treadmill work (5.6 km/hr, 2.5% grade; MR 190 kcal/m² · hr). Each value of mean skin temperature and sweating represents the average for two men during the second hour of an exposure; each value of heart rate and rectal temperature represents the average of the men at the end of the corresponding 2-hr exposure. The men were clothed in 5-oz. poplin suits. Data of Robinson (1).

by heart rates above 140, appeared only when the water vapor pressure in the air became so high that heat dissipation by evaporation was limited, and rectal and mean skin temperatures rose significantly above 38.5° and 36°C, respectively.

Heat Conductance

The burden of maintaining heat balance in the body in hot environments is borne to a great extent by the circulatory system. The conductance of heat from the tissues in which it is being produced to the skin and respiratory system where it is being dissipated is largely dependent on the circulation; simple conductance by body tissues alone would severely limit the rate of heat transfer from the interior to the surface of the body. This is particularly true in hot environments in which the skin temperature is elevated and the temperature gradient between core and surface of the body is reduced. The blood and tissue fluids, composed largely of water, have a high specific heat and high conductivity of heat; these properties are important in that large amounts of heat can be stored and transported with small changes in temperature.

We have found that a man who is well acclimatized to work in hot environments may maintain thermal equilibrium with a total cutaneous blood flow of 1.2 liters/m^2·min during a 6-hr period of work (MR 190 kcal/m^2·hr) in the heat (50°C with 18% RH) (2). In this situation about 20% of his total cardiac output of 11 liters/min is being circulated through the skin. The circulatory system of this man working in equilibrium with the hot environment is meeting at least three demands in addition to the requirements for ordinary metabolic processes in the resting man—(a) a large blood flow through the working muscles provides for respiratory exchange and conducts away the increased heat being produced there; (b) blood flow through the skin cools the blood and supplies the sweat glands with water at about 1.5 liters/hr; and (c) flow through the alimentary tract and liver transfers water and nutrients to sweat glands and working muscles. Higher total cutaneous blood flows than this probably cannot be maintained for long periods of time, although higher values may be observed in shorter exposures to more extreme heat, as illustrated in Fig. 2 and by the data of Eichna *et al.* (3).

The coefficient of heat conductance, expressed as kcal/m^2 · hr/°C difference between central and mean skin temperature, has been used as an index of total cutaneous blood flow by a number of investigators (4-9). Conductance is not considered to be an absolute measure of peripheral blood flow because of the complex variations of blood flow patterns and temperature gradients in the body, especially in the limbs. One complication in cool or cold environments is illustrated by the classic studies of Bazett (10-13) of the countercurrent exchanges of heat between the warm blood in the large arteries and the cool blood returning in the adjacent venae comitantes setting up longitudinal temperature gradients in the limbs. On the other hand, in warm and hot environments arterial blood is not precooled by heat exchange with venae comitantes. Venous blood returns through superficial veins and continues to be cooled in transit. In a warm man they found the temperature of blood in superficial veins of the finger to be 35.2°C and of the elbow 34.6°C (14). More recently Gisolfi and Robinson (15) found that blood temperature was 0.5°C higher in the superficial saphenous vein in the calf than in the same vein above the knee in men performing hard treadmill work in a hot room (40°C). The latter authors also presented evidence that a large fraction of the blood returning from the working muscles of the legs is shunted through penetrating veins into the superficial veins of the legs. These results thus confirm Bazett's suggestion that under hot conditions the returning venous blood flow in the limbs is shunted to the dilated superficial veins by constriction of the deep veins, including the venae comitantes. In spite of these complexities of the peripheral circulation in heat exchange, the coefficient of heat conductance is a useful indicator of total cutaneous blood flow, and Hertzman *et al.* (16) have found a high correlation between conductance and estimates of blood flow from local determinations with plethysmographic techniques.

Measurements of heat conductance have revealed many important variations in cutaneous blood flow (4-6, 8, 17, 18). In exposure of resting men to hot environments, maximal conductances of 35-50 kcal/m^2 · hr/°C have been reported by a number of investigators. Eichna *et al.* (3) found that thermal conductance and cutaneous blood flow during exposure were reduced with acclimatization in men exposed daily to a constant submaximal work-heat stress. Thus the cardiovascular strain was reduced as men's ability to regulate body temperature improved. On the other hand, Piwonka and Robinson (19) found in more severe work-heat stress that the capacity of men to circulate blood to the skin increased with acclimatization. Hellon *et al.* (20, 21) found that upon exposure to heat, older men (between the ages of 41 and 58) circulated 50% more blood to the forearm than did men 17-26 years old. The older men also had greater increases of cutaneous blood flow during work in the heat. Hardy *et al.* (7) and DuBois *et al.* (22) found that women have lower conductances than men in cool environments and higher conductances in warm environments.

Both the minimal tissue heat conductance in cold environments and the maximal rate of conductance in heat are increased during exercise by men, according to data of Robinson (2, 23). Figure 2 shows the conductances in a working man (MR 195 kcal/m^2 · hr) in a series of 90-min exposures to various environmental temperatures. Conductance during each period of observation is plotted against the subject's mean skin temperature, rectal temperature, gradient between rectal and mean skin temperature, and mean body temperature, respectively. In these experiments the stimulus that elicits the heat-regulatory response of cutaneous vasodilation and increases conductance could be represented by any one or any combination of these four measures of body temperature in steady states. In rapidly changing states the rectal temperature does not accurately represent the hypothalamic temperature, but in prolonged experiments the two should be closely related to each other and to the mean core temperature of the body. Heat conductance shows a close positive correlation with both mean skin temperature and mean body temperature, an inverse relation to the gradient between rectal and skin temperature, and no relation to rectal temperature. Minimal and maximal conductances of this working man were 20 in the cold (ET 9.5°C) and 130 in the hot (ET 35°C) environment. These values are more than twice as high as the corresponding maximal and minimal values that have been reported for resting subjects.

Figure 3 shows a direct comparison of conductances determined in another subject (LG) at rest and during work in a series of experiments performed in various effective temperatures ranging from 15° to 35°C (2). In the work

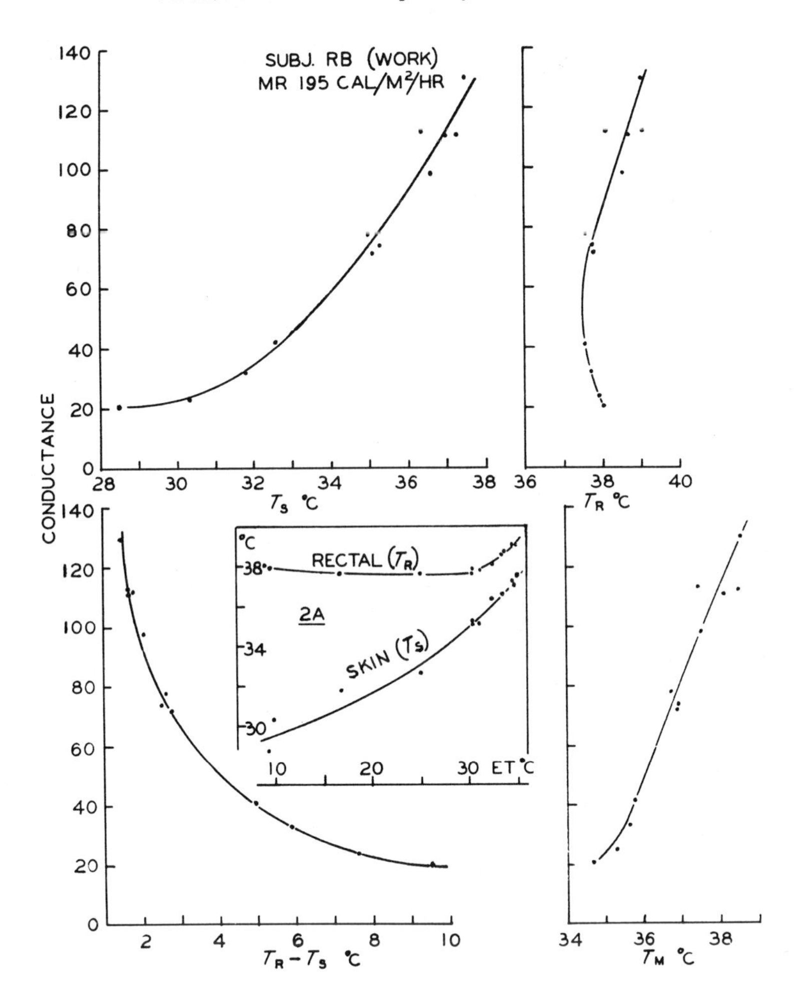

FIG. 2. Insert 2A shows rectal temperature (T_{re}) and mean skin temperature ($\bar{T}_s$) of subject RB clad in shorts, shoes, and socks during exposures to 12 different effective temperatures (ET 9.5° to 35°C) of the environment. The other graphs show, respectively, the relations of tissue heat conductance to $\bar{T}_s$ (mean skin temperature), $T_{re}- \bar{T}_s$, and to $\bar{T}_b$ (mean body temperature) of the subject in the 12 experiments. Effective temperature of the room was constant in each exposure, and work (5.6 km/hr, 2.5% grade) by the subject was the same in all cases. Each value plotted represents the average of measurements made in the last 60 min of a 90-min exposure. Data of Robinson (2).

experiments on subject LG, the relations of conductance to each of the four measures of body temperature are similar in character to those in subject RB (Fig. 2) over a corresponding range of environmental stresses. For the purpose of this discussion, the most important consideration of the data in Fig. 3 is the stimulating effect of work on conductance in the subject. In these steady-state experiments, conductance was much greater in work than in rest at any comparable measure of body temperature. It is particularly significant that at any given total body heat load ($\bar{T}_b$) the response was greater in work than in rest. In both men during work, conductance varied from 20 to 70 with no elevation of T_{re} above the control values observed in the cool environments. In the resting state during the 2-hr exposures to the cooler environments, the

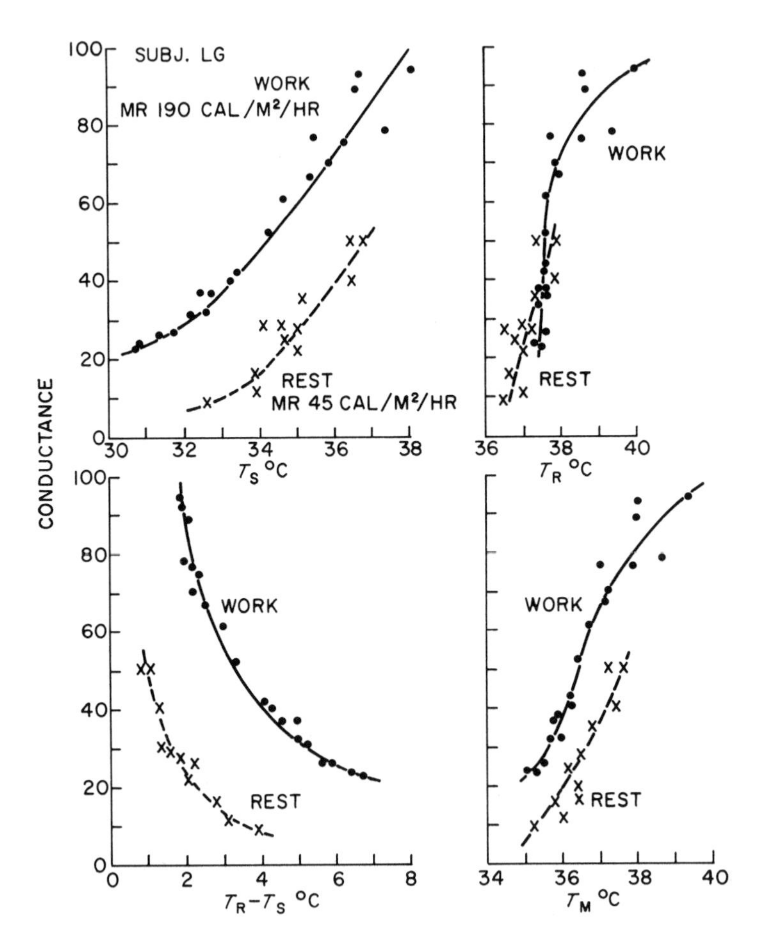

FIG. 3. The effect of work on the relation of tissue heat conductance to body temperature: T_{re} (rectal temperature), $\bar{T}_s$ (mean skin temperature), $T_{re} - \bar{T}_s$, and $\bar{T}_b$ (mean body temperature). Each value plotted represents the average of measurements made on subject LG, clad in a 5-oz. poplin suit, during the second hour of a 2-hr exposure. The effective temperature of the room was constant during each exposure, but varied from 15° to 35°C in the different 2-hr exposures. Data of Robinson (2).

lightly clad subject's rectal temperature tended to fall below control values observed in comfortable environments, and the data show a positive correlation between $\bar{T}_{re}$ and conductance.

The Effects of Work

The stimulating effect of neuromuscular work, which is summated with thermal stimuli to elicit the temperature regulatory responses, increased heat conductance, and cutaneous blood flow in this case cannot be fully explained on the basis of present knowledge. The author (2, 24, 25) has previously suggested the possibility of neuromuscular reflexes exciting the hypothalmic temperature regulatory center of a working man in a manner similar to reflexes that are known to excite respiration (26) during work. In fact, the relationships of the subject's heart rate (Fig. 4), sweat rate (24), and conductance (Fig. 3) to body temperature and to work in the above study are remarkably similar, and we have confirmatory data in identical experiments on four other subjects. It seems logical that the stimulus associated with work should be a thermal one, and, therefore, it might be an elevation of the temperature of special thermoreceptors in the muscles or in the veins draining warm blood from the working leg muscles, but we do not have conclusive evidence to support this hypothesis (15). Bazett (10,11) suggested that thermoreceptors located in the deeper cutaneous venous plexuses that drain the muscles may elicit temperature regulatory responses reflexly during work when the temperature of the blood from the muscles is elevated. Recent studies by Nielsen (27) failed to confirm this hypothesis.

Gisolfi and Robinson (15) found that at the beginning of hard treadmill work by men, the blood temperature in the superficial saphenous vein just below the knee rose sharply in the first 3 min and then followed a course parallel to the rise in femoral blood (deep vein) temperature, indicating that both veins were transporting blood directly from the working muscles of the foot and lower leg (Fig. 5A). Rhythmic rises and falls in temperature, synchronous with pace, were observed in the saphenous veins of all subjects; each rise in saphenous blood temperature corresponded with a contraction of the leg muscles in a step on the treadmill, providing further evidence that this superficial vein drains blood directly from the working muscles of the foot and lower leg (Fig. 5B). To test the interpretation suggested by the shifts in temperature, the concentration of blood lactate, an anaerobic metabolite, was used as a tracer to indicate the distribution of blood between the deep and superficial veins of the leg (15). The hypothesis was that, in hard leg work, blood lactate in the saphenous vein might lag behind that in the femoral, as it does in the arm veins, and then rise in early recovery. In very hard treadmill work lactate is produced principally in the leg muscles, and the delay in its distribution is ascribed to (a) its dilution in the

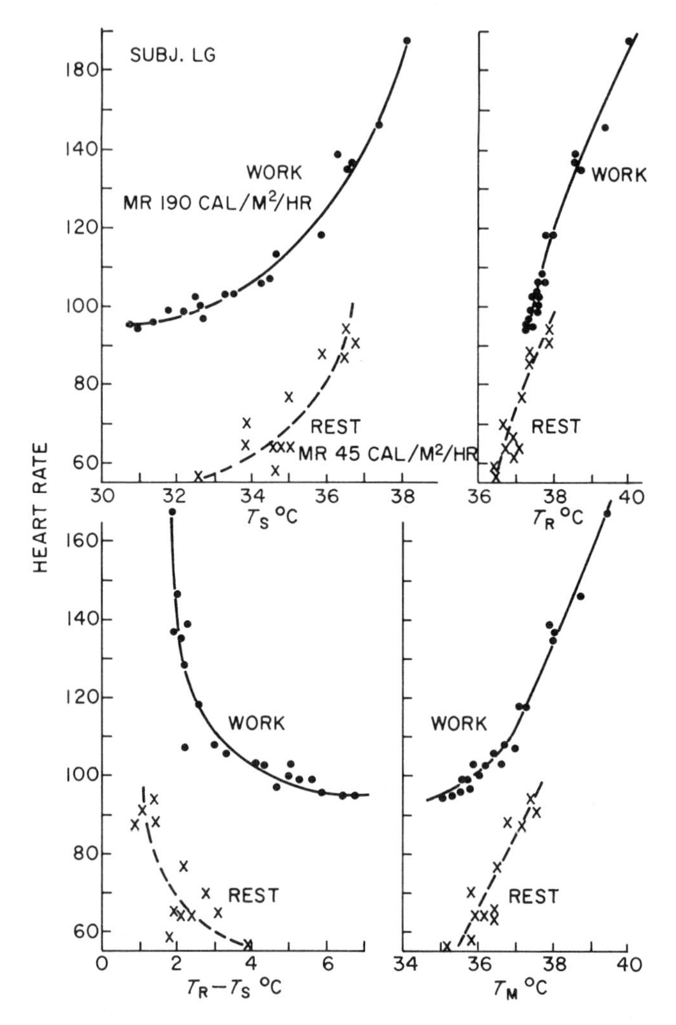

FIG. 4. Heart rate of subject LG in relation to body temperature during the 2-hr work and rest experiments described in Fig. 3.

heart by blood from less active tissues, (b) its diffusion from the blood into less active tissues, and (c) its diffusion from the plasma into the red blood cells. If the blood in the saphenous vein during hard leg work comes mainly from superficial tissues, then its rise in lactate concentration should lag behind that in the femoral vein as it does in the arm vein. On the other hand, if lactate concentration in the saphenous vein during hard work follows a course similar to that in the deep femoral vein, it would provide evidence that at least part of the saphenous flow comes directly from the working leg muscles.

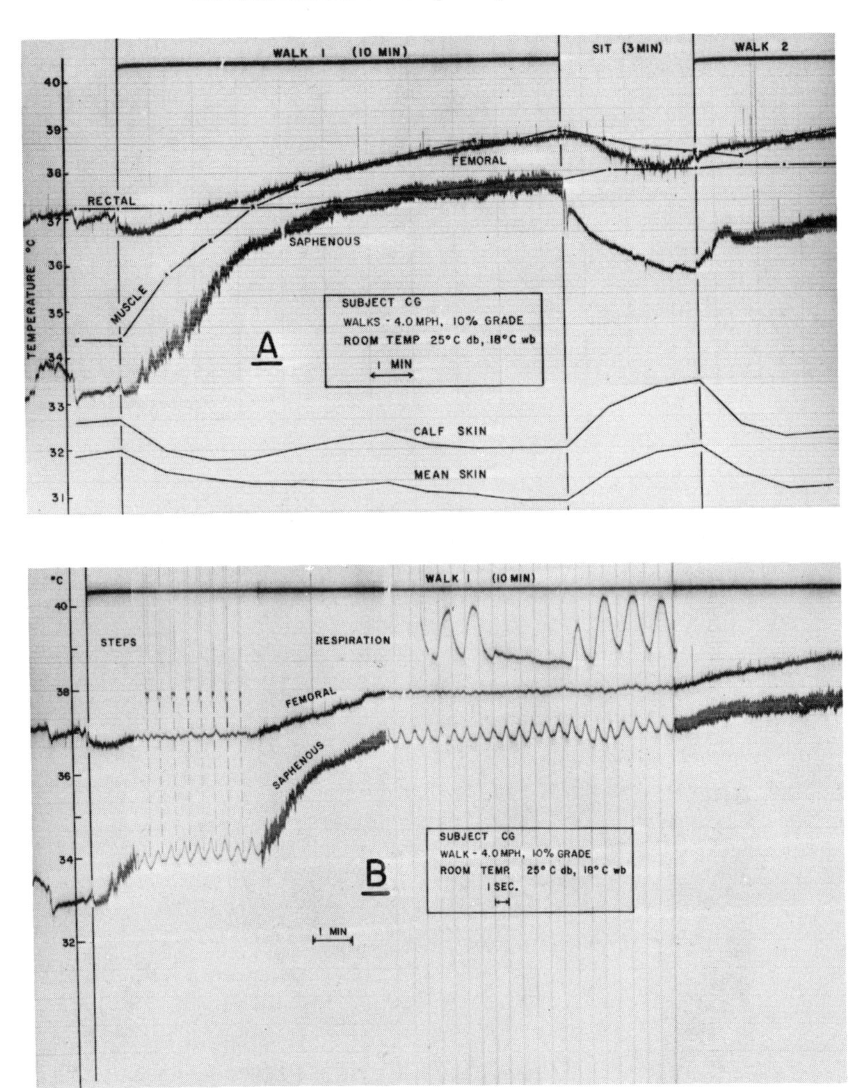

FIG. 5. (A) Relationships between intravascular, gastrocnemius muscle, rectal, and skin temperatures in subject CG during the first of a series of bouts of hard treadmill work. (B) High-speed recording during parts of walk 1 shown in A. The wide trace of saphenous vein blood temperature during work is due to a rhythmic rise and fall of Tsaph synchronous with the subject's steps on the treadmill. The rhythmic cycles were independent of the respiratory cycle, since they did not change when the subject held his breath. Both, walks 4.0 mph, 10% grade; room temp., 25°C, dry bulb; 18°C wet bulb. Data of Gisolfi and Robinson (15).

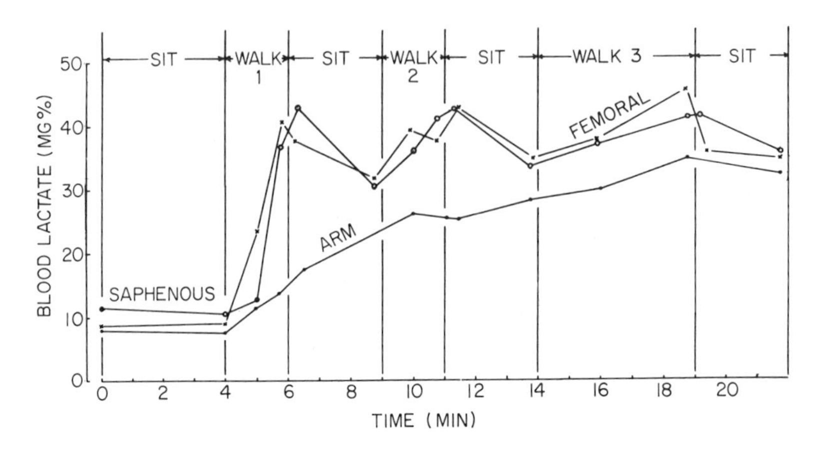

FIG. 6. Femoral, saphenous, and arm (cephalic) vein blood lactate concentrations of subject DL during alternating periods of hard treadmill work and rest in the warm room. (Walks, 4.5 mph, 9% grade; room temp., 40° C dry bulb; 23° C wet bulb.) Data of Gisolfi and Robinson (15).

Figure 6 shows femoral, saphenous, and cephalic (arm) vein blood lactate concentrations during hard treadmill work in the heat by subject DL. During work, lactate concentrations in femoral and saphenous veins rose together, more rapidly and to significantly higher values, than in the arm vein. These data confirm the temperature data and indicate that during work a large fraction of the blood in the saphenous vein at knee level is coming directly from working muscles of the foot and lower leg, even though the superficial veins are often constricted during hard work and it is difficult to sample blood from the saphenous through an indwelling catheter. During recovery, arm and leg blood lactate approach the same values as the lactate that was produced by the leg muscles during work is uniformly distributed in the body fluids (15).

The time relations of changes in both blood temperature and lactate concentrations in the deep and superficial veins of the leg indicate the function of communicating veins, probably of the indirect type, that allow blood to flow directly from the deep to superficial leg veins during hard treadmill exercise and recovery. This provides a more even venous return from the legs and contributes to temperature regulation by cooling the warm blood from the working leg muscles before it reenters the central circulation. The mechanisms providing for the operation of this perforating venous flow remain obscure. The rapidity with which the saphenous blood temperature changes occur with each step indicate that it is at least partially a passive effect of the rhythmically contracting muscles. The more gradual changes following the starts and stops of work probably depend on venomotor adjustments and the changes in heat production in the muscles. The return of blood through the superficial veins of the limbs during work undoubtedly contributes to the greatly increased coefficient of heat conductance observed in working men (Fig. 3).

Regulation of Cutaneous Blood Flow

There is a wide variation in blood flow to different skin regions in response to heat stimuli. Hertzman (28) and Hertzman and Dillon (29) have found the richness of blood supply to different skin areas to be arranged in descending order as follows: finger pad, ear lobe, toe pad, palm of hand, forehead, forearm, knee, and tibia. Within the hand, the rate of circulation in the digits is relatively greater than in the entire hand, and it is greater in the distal phalanges than in more proximal ones (18). It is now known that maximal rates of blood flow and the vasomotor regulation of flow to the hands and feet are radically different from those of the forearms and legs. Flow to hands and feet is increased in the heat largely by graded release of these vascular beds from vasoconstrictor tone (9, 30-32), whereas cutaneous flow to the forearm in response to heat is regulated largely by active vasodilator nerve impulses. That reflex vasodilation occurs in the fingers of the hand in response to heating skin areas remote from the hand itself has been known for many years, and it has been demonstrated by many investigators. Kerslake and Cooper (33) have proven that this reflex response to indirect heating must be due to excitation of cutaneous temperature receptors by demonstrating that its latent period is only 10-15 sec. Great and rapid changes in blood flow to the hands were observed by these workers in response to gentle radiant heating of the trunk or the legs. The response was proportional to the intensity of the stimulus. Snell (34) has demonstrated central excitation of the response by adding 2.4 kcal of heat to the body by intravenous infusion of a warm salt solution. The blood vessels of the hands are also sensitive to temperature directly, since in the reflexly dilated hand dilation is further increased by local heating of the hand (35, 36).

Evidence of active vasodilator control of blood flow to the periphery was presented by Grant and Holling (37) in 1938 when they observed that intense heating of the legs caused marked vasodilation of the forearms of men. A week or more after sympathetic ganglionectomy, heating failed to elicit the reflex response, indicating that in the intact man the response was brought about through active sympathetic vasodilator fibers to the forearm. Blocking cutaneous nerves to the forearm also prevented the cutaneous vasodilation of the arm in response to body heating. Active vasodilation is now known to participate in the regulation of flow to the skin of the forearm, as a result of work by Edholm *et al.* (38, 39) who also found that blocking cutaneous nerves to the forearm prevented cutaneous vasodilation in the anesthetized areas in response to body heating. During a period of body heating, anesthetizing the skin of the forearm reduced flow to the level observed in a cool environment.

The increase of blood flow to the forearm of a resting man in response to body heating is entirely cutaneous, with virtually no change in flow to the muscles. This is indicated by the work of Roddie *et al.* (40) who found that during body heating of resting men the oxygen saturation of venous blood returning from the forearm muscles was unchanged from values observed in a cool environment, but that the saturation of blood from the skin was doubled. The relationship is proven by recent experiments of Edholm and Cooper *et al.* (41-43) in which blood flow to the skin of the forearm was arrested by iontophoresis of adrenalin into the skin, and this was found to abolish the normal increase in flow to the forearm in response to body heating. Hertzman (44) has shown cutaneous vasodilation in the forearm during body heating by recording skin pulses. Barcroft *et al.* (45) observed a decrease in blood flow to the arm and leg muscles during body heating and an increase during cooling in resting men.

Several mechanisms may participate in the regulation of cutaneous blood flow in the forearm. First, in response to body heating there is a small increase occurring within seconds, which is due to reflex release of the vessels from vasoconstrictor tone, coinciding in time with dilation in the hand (32). Second, the principal response is delayed longer and is due to cholinergic vasodilator impulses, which may be blocked to a large extent by giving atropine (32) or by anesthetizing the cutaneous nerves (37). Delayed additional vasodilation, over and above the two vasomotor changes, may result from the direct effect of heat on the cutaneous vessels and possibly from the polypeptide bradykinin. Fox and Hilton (46) have found that bradykinin is formed in the skin during sweat gland activity and appears to participate in the cutaneous vascular response to heat.

Compensatory Adjustments

Cardiovascular strain, as evidenced by high heart rate and a high incidence of syncope, is one of the most prominent characteristics of the unacclimatized state in men exposed to severe heat stress, particularly if they attempt to perform work in the heat. Associated with the circulatory strain of unacclimatized men working in the heat are marked elevations of skin and rectal temperatures above control values. A dramatic improvement of cardiovascular stability takes place during acclimatization to work in the heat, and the improvement reaches virtual completion in 5 to 8 days (47-49). These circulatory changes in the subjects contribute to improvement in temperature regulation, as evidenced by reductions of both rectal and skin temperature, the former returning to control levels determined during work in a cool environment.

Circulatory stability could not be maintained with the large increments in cutaneous blood flow of men in response to heat stress without efficient compensatory adjustments in the circulation. The two principal adjustments that take place in this situation are (a) compensatory vasoconstriction in other vascular beds, shunting a greater fraction of the total cardiac output to the skin and (b) a 10-20% expansion of the circulating blood volume. An increment in heart output would also be advantageous in maintaining adequate circulation in the heat, but this does not occur consistently, and heart output of men in the erect posture even falls in the heat.

Evidence of compensatory vasoconstriction in the viscera of men exposed to heat is provided by measurements of renal blood flow. In severe heat (dry bulb 50°C; wet bulb 27°C), the blood flow to the kidneys of resting men may be reduced to 60-75% of control values observed in a cool environment (50-53). If the men exercise at a moderate rate (MR, 4 met) in the heat, renal blood flow is further reduced to 25-40% of control values observed for resting men in a cool environment. Dehydration (4-8% of body weight) superimposed upon work in the heat produces even greater reductions in renal blood flow (50). In the experiments of Radigan and Robinson (53) and of Smith *et al.* (50), when the subjects were not dehydrated they were in thermal equilibrium during work which they commonly performed for 6 hr in the heat. Cutaneous blood flow of men working in equilibrium under these conditions (dry bulb 50°C; wet bulb 27°C) was estimated from conductance values to be 1.2 liters/$m^2 \cdot$ min. The reductions in renal blood flow alone were estimated to be more than 50% of the total increase in blood flow to the skin under these conditions. More recently Rowell *et al.* (54) have found that hepatic blood flow of men is also substantially reduced during work in the heat. Compensatory vasoconstriction in other vascular beds probably also contributes to man's circulatory well-being in the heat.

Numerous workers have reported that the blood volumes of men increase above basal values during actual exposure to heat (55-61). Other workers have reported no change or actual reductions in blood volumes of men during acute exposures to heat or during acclimatization. Comparisons of circulating blood volumes of men with respect to summer and winter seasons have in some cases shown variable changes or no change with season (62-64) and, in others, significantly higher blood volumes in summer than in winter (61, 65). We (60) found that men well acclimatized to heat showed normal basal blood volumes during rest in a cool environment, and increases averaging 13% during exposure of 2-3 hr to moderate treadmill work in severe heat when water and electrolyte balance were maintained (60, 61). According to Bazett (57), acclimatization to heat produces even greater increases in blood volume than those indicated in short exposures. Bass *et al.* (56) found that, during continuous exposure of men to a room temperature of 49°C with 28% humidity, blood volume increased by an average of 15% on the fifth day; by the fourteenth day in the heat it had returned practically to the control level previously determined in a cool environment. The increase was due largely to expansion of the plasma volume. These workers found that expansion of plasma and thiocyanate spaces during heat acclimatization was made possible by retention of NaCl by sweat glands and kidneys in proportion to the increase in the volume of extracellular fluid.

Another factor that profoundly affects the circulatory adjustments of men to heat is the state of hydration of the body. Dehydration reduces heat tolerance of men, whether the loss of body water is due directly to failure to replace all of the water lost through sweat glands and kidneys or is secondary to NaCl deficiency resulting from failure to fully replace salt losses in the sweat and urine. Dehydration of normal men in hot environments reduces their plasma volumes under circumstances when they can least tolerate it. Adolph *et al.* (66) found in men dehydrated by 1-8% of body weight under desert conditions that reductions in circulating plasma volume amounted to 2½ times those which would be expected if water loss of the plasma were in proportion to the loss by the whole body. Myhre and Robinson (67) have recently confirmed this in men dehydrating rapidly during 2-3 hr of work in severe heat. They also found that if the men moved to a cool room and remained dehydrated for 4 hr longer, fluid was gradually shifted from other compartments into the plasma, replacing 50% of the loss in plasma volume. No further increase in plasma volume occurred when the men remained dehydrated for 19½ hr after leaving the heat. The wide variations in the changes of blood volume of men during exposure to heat that have been observed by different investigators probably depend on such factors as (a) acclimatization; (b) the states of hydration and electrolyte balance; and (c) the thermal states of the subjects during both the control and the experimental determinations.

From the discussion of the important role of an increased blood volume by men in maintaining an adequate circulation under conditions of heat stress and maximal cutaneous vasodilation, it is obvious that dehydration and its consequent reduction of plasma volume would increase the circulatory strain and reduce the tolerance of men to heat stresses. This effect was studied in great detail by Adolph and his colleagues (66) and by Pitts *et al.* (68). They found that water deficits as low as 1-2% of body weight caused measurable evidence of increased circulatory strain as indicated by increases in the heart rates and rectal temperatures of men resting or working in hot environments. The strain, under otherwise constant conditions of metabolic rate and heat stress, increased linearly with further increments in water deficit. Accompanying the increased heart rates of the men associated with dehydration were parallel increments in rectal temperature, indicating a definite failure of the circulation in its function of heat transfer from tissues to skin. Adolph called such failure *dehydration exhaustion*, when it is accompanied by characteristic symptoms of decreased work output, drowsiness, faintness, dyspnea, dry mouth, and restlessness.

A resting man who is acclimatized to a hot climate will maintain water balance accurately by drinking enough water to keep his thirst satisfied, and, because of this, the water content of the body is remarkably constant when measured each morning before the day's work begins. In contrast to this, Dill (69) and others (68, 70) have found that thirst does not always cause a working man to keep his water intake up to its rate of output. This is particularly true in unacclimatized men who may, in working in a hot environment, secrete large volumes of sweat containing relatively high concentrations of salt (40-70 mEq/liter) (69). In this case, sweat formation involves little modification of the osmotic pressure of the extracellular phase and, therefore, does not greatly reduce the water content of the cells; so it does not produce a degree of thirst proportional to the water deficit. On the other hand, an acclimatized man whose sweat is very dilute in salt (10-20 mEq/liter) is much more thirsty for a given water deficit and will come nearer to maintaining his water balance by voluntary drinking than the unacclimatized man. In this case, thirst is more intense because the loss of water raises the osmotic pressure of the extracellular fluid more and causes a withdrawal of fluid from the cells. Pitts *et al.* (68) and Adolph *et al.* (66) found that even acclimatized men working in the heat never voluntarily drink as much water as they sweat, even though this is advantageous for heat regulation, but usually drink at a rate of about 1/2 to 2/3 the water lost in sweat.

Serious dehydration, with consequent circulatory strain and hyperpyrexia, may result from this type of voluntary abstinence from water by working men, even when plenty of water is available to them. Adolph *et al.* (66) have termed it *voluntary dehydration* and found that it varies directly with the rate of sweating and the rate of work by the subject. Prevention of *voluntary dehydration* during work in the heat requires forced drinking of water, preferably water containing salt in the same concentration as that in the sweat the man is secreting.

Compensatory adjustments in the circulation of normal men in the heat may involve increments in cardiac output values measured in reclining subjects before acclimatization is complete, with a return to control values as they acclimatize (71). Scott *et al.* (71) observed small increases in heart output of reclining men during the first days of continuous exposure to a warm environment in the winter, but as they acclimatized their cardiac outputs in the heat returned to control values. Standing erect reduced cardiac output of unacclimatized men in the heat, but after several days of acclimatization posture made little difference in cardiac output. Similarly, Asmussen (72) found no significant difference in the reclining cardiac index of men between winter in Boston and Copenhagen and the subtropical summer climate of the Mississippi Delta after 10-30 days of acclimatization. Burch *et al.* (73, 74) observed significant increments in heart output of reclining patients in acute changes from air-conditioned hospital wards to the warm, humid environment of New Orleans in summer.

In the Mississippi study Asmussen (72) found that cardiac outputs during the first ½ hr of hard treadmill work (M 7 met) in humid heat were the same as in control walks in the winter in partially acclimatized men. However, as the work in the heat was continued, cardiac output declined, heart rate increased markedly, stroke volume fell, and marked elevations of rectal temperature occurred in the third and fourth half hours. Dill *et al.* (75) observed a similar failure of circulation in men working in a hot environment.

Respiratory Reactions to Heat

Hyperthermia is known to produce hyperpnea and hypocapnia in resting men (76-79) and to exaggerate respiratory responses to increased CO_2 tension (77, 80). Bazett (76) and Landis (79) studied hyperpnea of men in hot baths where respiration may be deep and slow resulting in marked reductions in alveolar and arterial p^{CO_2}. The disturbing effects of hypocapnia on the arterial blood p^{CO_2} and pH of a resting man may more than offset the advantages of the small increase in evaporative cooling which he gains by hyperventilation during exposure to heat. Thus the respiratory responses of man, who cools his skin by the evaporation of sweat, differs from nonsweating mammals in which the chief mechanism for increasing evaporative cooling in the heat is by aerating the wet air passages by rapid shallow panting.

Since hyperthermia in the resting man may result in hypocapnia, it is of interest also to determine the role of the hyperthermia of work on respiration and its regulation, especially during work in hot environments. Bannister *et al.* (81) and Cotes (82) have proposed that the increased central temperature produced by work participates in the regulation of respiration. DeJours *et al.* (83) have found that increments up to 1°C in central body temperature in exercise do not produce relative hyperventilation. Ts'ao *et al.* (84) studied the relations of men's pulmonary ventilations to rectal temperature, mean skin temperature, and mean body temperature in the last 15 min of 2-hr periods of treadmill work (M 190 kcal/m^2 · hr) in each of the 18 different environments indicated in Figs. 1 and 2. The relationships are presented graphically in Fig. 7, and it is clear from the data that the men's ventilation while performing this constant rate of work was unaffected by variations of skin temperature between 31° and 36°C, of rectal temperature between 37.5° and 38.5°C and of mean body temperature between 35.6° and 37.3°C. The men's body temperature did not exceed these limits in the 2-hr work periods in room temperatures up to 50°C with 18% relative humidity or 38°C with 65% humidity (Fig. 1). Hyperventilation, as indicated by increments of ventilation per liter of O_2 consumption, occurred only in the most extreme heat stresses when rectal temperature exceeded 39°C, and mean body temperature was above 38°C.

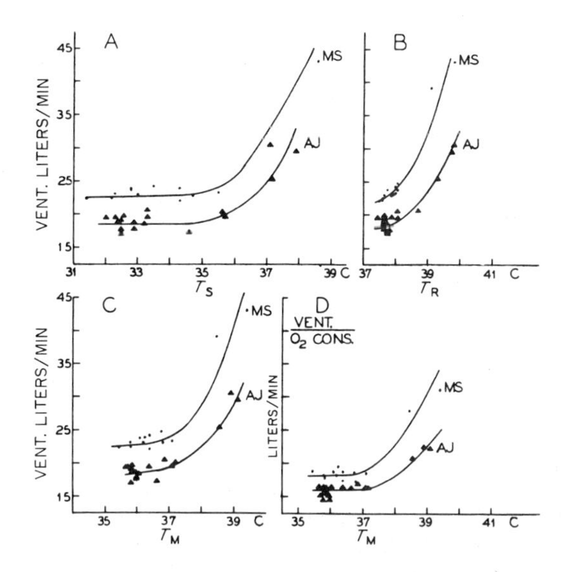

FIG. 7. The relations of ventilation (STPD) to mean skin (T_s) A, rectal (T_r) B, and mean body (T_m) C temperatures of subjects AJ and MS. In D, ventilation per liter of O_2 consumption is plotted against T_b. The values were measured in the last 15 min of 2-hr work experiments in which the effective temperature in the room was varied (15-35° C) from experiment to experiment. The work (5.6 km/hr, 2.5% grade) and the environment were constant in each experiment. Data of Ts'ao et al. *(84).*

References

1. Robinson, S. (1949). *In* "Physiology of Heat Regulation and the Science of Clothing" (L. H. Newburgh, ed.), Chapter 11, p. 340. Saunders, Philadelphia, Pennsylvania.
2. Robinson, S. (1963). Circulatory adjustments of men in hot environments. *Temp. Meas. and Contr. in Sci. and Ind., Proc. Symp., 4th, 1961*, Vol. 3, Part 3, 287, Reinhold Publ., New York.
3. Eichna, L. W., Park, C. R., Nelson, N., Horvath, S. M., and Palmes, E. D. (1950). Thermal regulation during acclimatization to hot, dry environment. *Amer. J. Physiol.* **163**, 585.
4. Burton, A. C. (1934). The application of theory of heat flow to the study of energy metabolism. *J. Nutr.* **5**, 497.
5. Burton, A. C., and Bazett. H. C. (1936). A study of the average temperature of the tissues, of the exchanges of heat and vasomotor responses in man by means of a bath calorimeter. *Amer. J. Physiol.* **117**, 36.
6. Hardy, J. D. (1937). The physical laws of heat loss from the human body. *Proc. Nat. Acad. Sci. U. S.* **12**, 631.
7. Hardy, J. D., Milhorat, A. T., and DuBois, E. F. (1941). Basal metabolism and heat loss

of young women at temperatures from 22° to 35°C. *J. Nutr.* **21**, 383.
8. LeFever, J. (1911). "La Chaleur Animale et Bioenergétiques." Masson, Paris.
9. Warren, J. V., Walter, C. W., Romano, J., and Stead, E. A., Jr. (1942). Blood flow in the hand and forearm after paravertebral block of the sympathetic ganglia. *J. Clin. Invest.* **21**, 665.
10. Bazett, H. C. (1949). *In* "Physiology of Heat Regulation and the Science of Clothing" (L. H. Newburgh, ed.), Chapter 4, p. 147. Saunders, Philadelphia, Pennsylvania.
11. Bazett, H. C. (1951). Theory of reflex controls of body temperature in rest and exercise. *J. Appl. Physiol.* **4**, 245.
12. Bazett, H. C., Mendelson, E. S., Love, L., and Libet, B. (1948). Precooling of blood in the arteries, effective heat capacity and evaporative cooling of the extremities. *J. Appl. Physiol.* **1**, 169.
13. Bazett, H. C., Love, M., Newton, M., Eisenberg, L., Day, R., and Forster, R., II. (1948). Temperature changes in blood flowing in arteries and veins in man. *J. Appl. Physiol.* **1**, 3.
14. Bazett, H. C., and McGlone, B. (1927). Temperature gradients in the tissues of man. *Amer. J. Physiol.* **82**, 415.
15. Gisolfi, C., and Robinson, S. (1970). Venous blood distribution in the legs during intermittent treadmill work. *J. Appl. Physiol.* **29**, 368.
16. Hertzman, A. B., Randall, W. C., Edestrom, H. E., and Peiss, C. N. (1951). The cutaneous vascular responses to rising environmental temperature. WADC Tech. Rep. No. 6680, Part 6.
17. Hardy, J. D., and Soderstrom, G. F. (1938). Heat loss from the nude body and peripheral blood flow at temperatures of 22° to 31°C. *J. Nutr.* **16**, 493.
18. Hertzman, A. B., Randall, W. C., and Jochim, K. E. (1946). Estimation of cutaneous blood flow by the photoelectric plethysmograph. *Amer. J. Physiol.* **145**, 716.
19. Piwonka, R. W., and Robinson, S. (1967). Acclimatization of highly trained men to work in severe heat. *J. Appl. Physiol.* **22**, 9.
20. Hellon, R. F., and Lind, A. R. (1958). The influence of age on peripheral vasodilation in a hot environment. *J. Physiol. (London)* **141**, 262.
21. Hellon, R. F., and Clarke, R. S. J. (1959). Changes in forearm blood flow with age. *Clin. Sci.* **18**, 1.
22. DuBois, E. F., Ebaugh, F. G., Jr., and Hardy, J. D. (1952). Basal heat production and elimination of thirteen normal women at temperatures of 22 to 35°C. *J. Nutr.* **48**, 257.
23. Robinson, S. (1963). Temperature regulation in exercise. *Pediatrics* **32**, 691.
24. Robinson, S. (1962). The regulation of sweating in exercise. *Advan. Biol. Skin* **3**, 152.
25. Gisolfi, C., and Robinson, S. (1970). Central and peripheral stimuli regulating sweating during intermittent work in men. *J. Appl. Physiol.* **29**, 761.
26. Comroe, J. H., Jr., and Schmidt, C. F. (1943). Reflexes from the limbs as a factor in the hyperpnea of muscular exercise. *Amer. J. Physiol.* **138**, 536.
27. Nielsen, B. (1969). Thermoregulation in rest and exercise. *Acta Physiol. Scand., Suppl.* **323**, 24.
28. Hertzman, A. B. (1938). The blood supply of various skin areas as estimated by the photoelectric plethysmograph. *Amer. J. Physiol.* **124**, 328.
29. Hertzman, A. B., and Dillon, J. B. (1939). Selective vascular reaction patterns in the nasal septum and skin of the extremities and head. *Amer. J. Physiol.* **127**, 671.
30. Freeman, N. E. (1935). Effect of temperature on rate of blood flow in the normal and sympathectomized hand. *Amer. J. Physiol.* **113**, 385.
31. Roddie, I. C., Shepherd, J. T., and Whelan, R. F. (1957). A comparison of heat elimination from the normal and nerve-blocked finger during body heating. *J. Physiol. (London)* **138**, 445.

32. Roddie, I. C., Shepherd, J. T., and Whelan, R. F. (1957). Contribution of constrictor and dilator nerves to skin vasodilation during body heating. *J. Physiol. (London)* **136**, 489.
33. Kerslake, D. McK., and Cooper, K. E. (1950). Vasodilation in the hand in response to heating the skin elsewhere. *Clin. Sci.* **9**, 31.
34. Snell, E. S. (1954). Relationship between vasomotor response in the hand and heat changes in the body induced by intravenous infusions of hot or cold saline. *J. Physiol. (London)* **125**, 361.
35. Roddie, I. C., and Shepherd, J. T. (1956). Blood flow through the hand during local heating, release of sympathetic vasomotor tone by indirect heating and a combination of both. *J. Physiol. (London)* **131**, 657.
36. Allwood, M. J., and Burg, H. S. (1954). Effect of local temperature on blood flow in the human foot. *J. Physiol. (London)* **124**, 345.
37. Grant, R. T., and Holling, H. E. (1938). Vascular responses of the human limb to body warming. *Clin. Sci.* **3**, 273.
38. Edholm, O. G., Fox, R. H., and MacPherson, F. K. (1956). The effect of cutaneous anaesthesia on skin blood flow. *J. Physiol. (London)* **132**, 15P.
39. Edholm, O. G., Fox, R. H., and MacPherson, R. K. (1957). Vasomotor control of the cutaneous blood vessels in the human forearm. *J. Physiol. (London)* **139**, 455.
40. Roddie, I. C., Shepherd, J. T., and Whelan, R. F. (1956). Evidence from venous oxygen saturation measurements that the increase in forearm blood flow during body heating is confined to the skin. *J. Physiol. (London)* **134**, 444.
41. Cooper, K. E., Edholm, O. G., Fletcher, J. G., Fox, R. H., and MacPherson, R. K. (1954). Vasodilation in the forearm during indirect heating. *J. Physiol. (London)* **125**, 57P.
42. Cooper, K. E., Edholm, O. G., and Mottram, R. F. (1955). Blood flow in skin and muscle of the human forearm. *J. Physiol. (London)* **128**, 258.
43. Edholm, O. G., Fox, R. H., and MacPherson, R. K. (1956). The effect of body heating on the circulation in skin and muscle. *J. Physiol. (London)* **134**, 612.
44. Hertzman, A. B. (1953). Some relations between skin temperature and blood flow. *Amer. J. Phys. Med.* **23**, 233.
45. Barcroft, H. K., Bock, D., Hensel, H., and Kitchin, A. H. (1955). Die Muskeldurchblutung des Menschen bei indirekter Erwarmung und Abkulung. *Pfluegers Arch. Gesamte Physiol. Menschen Tiere* **261**, 199.
46. Fox, R. H., and Hilton, S. M. (1958). Bradykinin formation in human skin as a factor in heat vasodilation. *J. Physiol. (London)* **142**, 219.
47. Bean, W. B., and Eichna, L. W. (1943). Performance in relation to environmental temperature. *Fed. Proc., Fed. Amer. Soc. Exp. Biol.* **2**, 144.
48. Robinson, S., Turrell, E. S., Belding, H. S., and Horvath, S. M. (1943). Rapid acclimatization of men to work in hot climates. *Amer. J. Physiol.* **140**, 168.
49. Taylor, H. L., Henschel, A. F., and Keys, A. (1943). Cardiovascular adjustments of man in rest and work during exposure to dry heat. *Amer. J. Physiol.* **139**, 583.
50. Smith, J. H., Robinson, S., and Pearch, M. (1952). Renal responses to exercise, heat and dehydration. *J. Appl. Physiol.* **4**, 659.
51. Kenney, R. A. (1952). Effect of exercise in hot, humid environments on renal function of West Africans. *J. Physiol. (London)* **118**, 26P.
52. Kenney, R. A. (1952). Effect of hot, humid environments on renal function of West Africans. *J. Physiol. (London)* **118**, 25P.
53. Radigan, L. R., and Robinson, S. (1949). Effects of environmental heat stress and exercise on renal blood flow and filtration rate. *J. Appl. Physiol.* **2**, 185.

54. Rowell, L. B., Blackmon, J. R., Martin, R. H., Mazzarelli, J. A., and Bruce, R. A. (1965). Hepatic clearance of indocyanine green in man under thermal and exercise stresses. *J. Appl. Physiol.* **20**, 384.
55. Barcroft, J., Binger, C. A., Bock, A. V., Doggert, J. H., Forbes, H. S., Harrop, G., Meakins, J. C., and Redfield, A. C. (1922). The effect of high altitude on physiological processes of the human body. *Phil. Trans. Roy. Soc. London, Ser. B* **211**, 351.
56. Bass, D. E., Kleeman, C. R., Quinn, M., Henschel, A., and Hegnauer, A. H. (1955). Mechanisms of acclimatization to heat in man. *Medicine (Baltimore)* **34**, 323.
57. Bazett, H. C., Sunderman, F. W., Doupe, J., and Scott, J. C. (1940). Climatic effects on volume and composition of the blood in man. *Amer. J. Physiol.* **129**, 69.
58. Conley, C. L., and Nickerson, J. L. (1945). Effects of temperature changes on the water balance of man. *Amer. J. Physiol.* **143**, 373.
59. Glickman, N., Hick, L. K., Keeton, R. W., and Montgomery, M. M. (1941). Blood volume changes in men exposed to hot environmental conditions for a few hours. *Amer. J. Physiol.* **134**, 165.
60. Robinson, S., Kincaid, R. K., and Rhamy, R. K. (1950). Effects of desoxycorticosterone acetate on acclimatization of men to heat. *J. Appl. Physiol.* **2**, 399.
61. Sunderman, F. W., Scott, J. C., and Bazett, H. C. (1938). Temperature effects on serum volume. *Amer. J. Physiol.* **132**, 199.
62. Forbes, W. H., Dill, D. B., and Hall, F. G. (1940). Effect of climate upon the volumes of blood and tissue fluid in man. *Amer. J. Physiol.* **130**, 739.
63. Gibson, J. G., Jr., and Evans, W. A., Jr. (1937). Clinical studies of blood volume. *J. Clin. Invest.* **16**, 301.
64. Talbott, J. H., Edwards, H. T., Dill, D. B., and Drastich, L. (1933). Physiological responses to high environmental temperatures. *Amer. J. Trop. Med.* **13**, 381.
65. Maxfield, M. E., Bazett, H. C., and Chambers, C. C. (1941). Seasonal and postural changes in blood volume. *Amer. J. Physiol.* **133**, 128.
66. Adolph, E. F. *et al.* (1947). "Physiology of Man in the Desert," Chapters X-XVI. Wiley (Interscience), New York.
67. Myhre, L. G., and Robinson, S. (1971). Plasma volume during and following acute dehydration by exposure to environmental and work stresses. *Int. Congr. Physiol. Sci. Proc. 25th, 1971 Abstract, p. 411.*
68. Pitts, G. C., Johnson, R. E., and Consolazio, F. C. (1944). Work in the heat as affected intake of water, salt and glucose. *Amer. J. Physiol.* **142**, 253.
69. Dill, D. B. (1938). "Life, Heat and Altitude." Harvard Univ. Press, Cambridge, Massachusetts.
70. Adolph, E. F., and Dill, D. B. (1938). Observations on water metabolism in the desert. *Amer. J. Physiol.* **123**, 369.
71. Scott, J. C., Bazett, H. C., and Mackie, G. C. (1940). Climatic effects on cardiac output and the circulation in man. *Amer. J. Physiol.* **139**, 102.
72. Asmussen, E. (1940). Cardiac output in rest and work in humid heat. *Amer. J. Physiol.* **131**, 54.
73. Burch, G. E., and Hyman, A. (1959). Influence of tropical weather on output of volume, work, and power by the right and left ventricles of man at rest in bed. *Amer. Heart J.* **57**, 247.
74. Burch, G. E., DePasquale, N., Hyman, A., and DeGraff, A. C. (1959). Influence of tropical weather on cardiac output, work and power of right and left ventricles of man resting in hospital. *Arch. Intern. Med.* **104**, 553.
75. Dill, D. B., Edwards, H. T., Bauer, P. S., and Levenson, E. J. (1931). Physical performance in relation to external temperature. *Arbeitsphysiologie* **4**, 508.

76. Bazett, H. C. (1927). Physiological responses to heat. *Physiol. Rev.* **7**, 531.
77. Cunningham, D. J. C., and O'Roirdan, J. L. H. (1957). The effect of a rise in temperature of the body on the respiratory response to carbon dioxide at rest. *Quart. J. Exp. Physiol. Cog. Med. Sci.* **42**, 229.
78. Hill, L., and Flack, M. (1909). The influence of hot baths on pulse frequency, blood pressure, body temperature, breathing volume, and alveolar tension in man. *J. Physiol. (London)* **38**, 57P.
79. Landis, E. M., Long, W. L., Dunn, J. W., Jackson, C. L., and Meyer, U. (1926). Studies on the effects of baths on man; effects of hot baths on respiration, blood and urine. *Amer. J. Physiol.* **76**, 35.
80. Lambertsen, C. J. (1968). *In* "Medical Physiology" (V. B. Mountcastle, ed.), pp. 764-766. Mosby, St. Louis, Missouri.
81. Bannister, R. G., Cunningham, D. J. C., and Douglas, C. G. (1954). The CO_2 stimulus to breathing in severe exercise. *J. Physiol. (London)* **125**, 90.
82. Cotes, J. E. (1955). The role of body temperature in controlling body ventilation in exercise. *J. Physiol. (London)* **129**, 554.
83. DeJours, P., Teillac, A., Girand, F., and Lacaisse, A. (1958). Etude du role de hyperthermie centrale modérée dans la regulation de la ventilation de l'exercise musculaire chez l'homme. *Rev. Fr. Etud. Clin. Biol.* **3**, 755.
84. Ts'ao, C. H., Meyer, F. R., Epperson, B. E., and Holgersen, L. O. (1962). The influence of temperature on respiration in exercising men. *Physiologist* **5**, 222.

VII

Some Nutritional and Metabolic Aspects of Exposure to Heat

R. E. JOHNSON

Introduction

One way to order the priorities among nutrients is in terms of the rapidity with which disabling syndromes may appear in primary deficiency, that is, partial or total deprivation in healthy working men. By this set of criteria, the time span for oxygen is minutes, water hours, minerals and total calories a few days; protein and vitamins 2 weeks or so. It is not remarkable that water, salt, and calories have been subjects of continuing intensive interest and research for students of man in the desert.

This chapter will deal with two studies we have made in the past 2 years, the first the water economy in prolonged hard work in the heat and the second a reassessment of the specific dynamic action of fat in relation to physical work and a hot environment. These theoretical questions were two of many that arose during the studies Sargent and I did for the United States Air Force on the all-purpose survival ration for hot, cold, or temperate environments. The practical results of those studies were summarized in military reports (1). Some of the theoretical aspects specifically relating nutrition to the thermal environment were published in journal articles (2).

Water and Osmotic Balances and Performance

The general conclusion of Sargent and Johnson (1) was that there is such a thing as an all-purpose survival ration, useful in all environments, and based on the broad concept of the balanced diet. This states that there is an optimal ratio of calories from protein to carbohydrate and fat that, given adequate osmotic intake and vitamins, minimizes clinical deterioration under conditions of calorie deprivation, water deprivation and the stresses of work and environmental exposure. Wide deviations from this balance cause clinical, physiological, or biochemical deviations in the direction of disturbed function or actual disease.

The beneficial effects of adequate water intake during work in the heat were quantified early in World War II. In a study by Pitts *et al.* (3), a tough, well-motivated young soldier, in good training and specifically acclimatized for the task, walked at 39°C. In successive experiments he took no water, water *ad lib,* or water equal to the weight loss (Fig. 1). Without water he became hyperthermic. With complete water replacement he remained normothermic. Thus, the beneficial effects of acclimatization were suppressed by hypohydration. Therefore, acclimatization in one aspect is a function of the water balance. Adolph *et al.* (4) expanded on the control of water economy, and Gamble (5) emphasized the osmotic aspect. We feel that some of the ill effects of heat are produced by primary water deficiency, as shown in Fig. 1, or by secondary water deficiency. A negative osmotic balance carries with it an obligatory loss of water and, therefore, susceptibility to heat disorder. In salt deficiency there is

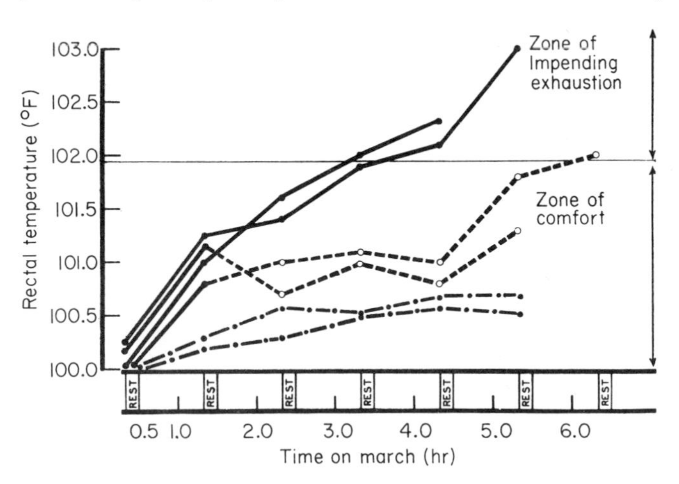

FIG. 1. Rectal temperature in subject JS walking at 6 km/hr on a 2.5% grade, at 39°C and 35% RH. –, Water loss not replaced; ––, water drunk ad lib*; –·–, Water loss replaced hourly. From Pitts* et al. *(3).*

susceptibility not only to cramp but to hyperthermia. [We feel also that Gamble (5) made an error in logic. He tested carbohydrate against starvation, and showed convincingly that carbohydrate was better with respect to water balance than starvation. He then concluded that carbohydrate is the best possible survival ration, without ever testing other combinations of nutrients.]

I have always been puzzled by an apparent paradox in Dill's (6) demonstration that acclimatization is accompanied by an increase in the rate of sweating for a fixed task, marching at a fixed pace for a fixed distance. From the standpoint of survival value, it might be supposed that acclimatization would lead not to an increase in the rate of water loss, but to a decrease and a net saving of water. We have addressed outselves to this paradox in a study of long-distance runners (7).

My colleagues in this study were Kachadorian, Serbin, Gerber, Buffington, Bailey, and Wakat from the University of Illinois, and Woodall from Eastern Illinois University. Each July we organize a 20-km road race, to and from the Lincoln Log Cabin State Memorial. In 1970, we made measurements on 136 men who finished at times from 66 to 120 min. The weather was warm, sunny, and humid. We measured the body weight before and after the race, collected urine before and after the race, and tabulated all food and fluid ingested between weighings.

The accurate measurement of the water balance under field conditions is not easy. To arrive at a simple formulation, let me review some well-known equations. Sanctorius (8) formulated the nutrient balance equation

$$\text{Balance} = \text{input} - \text{output} \tag{1}$$

This equation contains implicitly the law of conservation of mass, and, when used for thermal balance, the law of conservation of energy. These laws were not generally recognized until 250 years later.

Peters (9) and later Passmore *et al.* (10) have used a three-phase version of the Sanctorian equation to calculate the water balance. Consolazio *et al.* (11) called this the Peters-Passmore equation, in which the water balance is calculated without water intake or output.

$$\text{Water balance} = (\text{weight } 2 - \text{weight } 1) + (\text{mass } CO_2 \text{ produced} - \text{mass } O_2 \text{ consumed}) + (\text{solids excreted} - \text{solids consumed}) + H_2O \text{ oxidation} \tag{2}$$

The most difficult features of Eq. (2) are the H_2O oxidation, which has to be calculated from the metabolic mixture, and the solids excreted, part of which is in the sweat and not collectable.

Our simplified formulation rests on a number of features of the road race. The body weights before and after the race are known. No urine or feces are passed between weighings. The respiratory quotient is near 0.75, so that mass

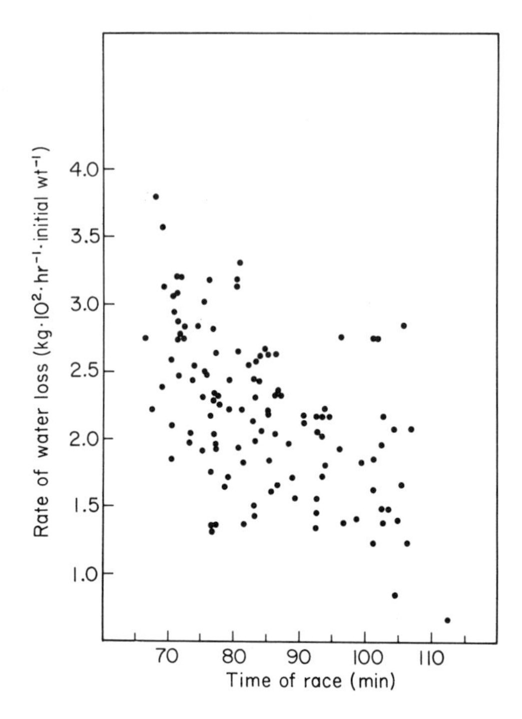

FIG. 2. Correlation between time in a 20-km road race in Charleston, Illinois (July 1970) and water loss expressed in terms of grams per kilogram body weight-race time. The total number finishing and studied was 119. From Johnson et al. *(7).*

CO_2 = mass O_2. No solids are consumed and the only solids excreted are in the sweat, and it is so dilute that these solids can be neglected. The source of fuel is predominantly fat, and so the H_2O oxidation is about 0.5 gm/liter O_2. The race time is known to the closest second. The $\dot{V}O_2$ is about 50 ml/kg · min.

$$\underset{(\mathrm{gm})}{\text{Water balance}} = (\underset{(\mathrm{gm})}{\text{weight 2}} - \text{weight 1}) + \left(\frac{25}{1000} \times \underset{(\mathrm{min})}{\text{Race time}} \times \underset{(\mathrm{kg})}{\text{weight 1}}\right) \quad (3)$$

We are interested in the practical as well as the theoretical aspects of work physiology. There was an excellent correlation between performance and the rate of water loss per kilogram weight and minutes of race time (Fig. 2). We have shown also that in studies of the water economy in prolonged hard work the water of oxidation can account for as much as 250 ml of water put into the system. This is not a negligible amount.

Our data agree with those of Pugh *et al.* (12) for marathon runners. They found that even under cool conditions the body temperature was up to 40°C, and the water loss was an average 1.8 kg. They suggested that marathon runners were in part displaying the characteristics of heat acclimatization.

We can now resolve the paradox of increased sweating with acclimatization. Our men who could sweat the fastest also ran the 20-km race the fastest. For escape or catching food this kind of excellence might pay off in survival.

The Heat Increment of Fat

The heat increment of food eaten (also called specific dynamic action by Rubner, commonly abbreviated SDA) has always occupied a rather obscure position in human nutrition. These calories are sometimes referred to as "waste heat," and have been troublesome to the work physiologist because, according to tradition, they cannot be used for muscular work. Mitchell (13) has given us a good review of heat increment in man and animals as well as some important experimental studies (14). He and his colleagues demonstrated clearly that fat lends more protection against cold stress than do isocaloric amounts of protein or carbohydrate.

Most of the research on heat increment has been on that of protein (e.g., Horvath's group, 15), probably for two main reasons. First, weight-for-weight protein has the largest SDA of the four energy-providing nutrients, and, second, it can contribute to thermal balance in a way that could be important in exposure to cold. The literature on the heat increment of fat is sparse, and there have been remarkably few studies on the interaction between fat ingestion, exercise, and thermal environment.

We have been interested in fat as a nutrient for many years. There are both theoretical and practical reasons. First, the major source of energy for prolonged moderate work is ultimately fat, as studies of the metabolic mixture show (16). The post-exercise ketosis of Courtice and Douglas (17) is one dramatic aspect of fat metabolism (18). Second, the history of survival rations has had a continuing theme that a high-fat diet is good because of its high caloric density. The literature on pemmican makes remarkably interesting reading (19). Finally, both obesity and atherosclerosis seem to be related to the amounts and kinds of lipids in the customary diet. My colleagues D. Bailey, Daria Harry, and Ingrid Kupprat addressed themselves to the heat increment of fat in an important but neglected aspect, namely recovery after prolonged moderate work.

Our original plan had been to study the metabolic mixture after a high-fat meal before a 1-hr walk, during a 1-hr walk, and in 2 hr of rest subsequently. Three different temperatures were to be imposed during and after the walk, warm temperate, or cool. So far we are still in the temperate part of the study.

The subjects were both men and women, neither particularly young nor particularly fit. They came to the laboratory postabsorptive for at least 12 hr. There were four possible combinations of eating or working. In all experiments there was a 1-hr rest period followed by 1 hr of "no-eat" or "eat" 750 kcal of fat, followed by 1 hr of "no-walk" or "walk" at 5 km/hr on a level treadmill. In all experiments the last 2 hr were "recovery 1" and "recovery 2" without food but with water *ad lib.* Respiratory metabolism was measured every 15 min by the method of Johnson *et al.* (20). Urine was collected each hour for the estimation of total nitrogen. The metabolic mixture was calculated as described by Consolazio *et al.* (11).

In the first place, we showed that the SDA of a high-protein meal could be detected (Fig. 3) (20a). JS was the subject, a traumatic paraplegic student who can and does sit quietly for hours. In the second place, we found that under our conditions there was a progressive rise in the O_2 consumption during 5 hr at rest without food and without exercise (Fig. 4) (20a) and at the same time a 90-min cycle amounting to as much as 10% from trough to peak. Clearly, such variations complicate the whole picture, because the total SDA of fat is only about 5% of ingested calories. We substantiated the study of Buskirk *et al.* (21), who showed considerable variations from time to time in the daily life of a soldier. Meals, with their own SDA, complicated that study. The "basal" metabolism is surely not constant.

With all the best controls we could devise, we finally could regularly and

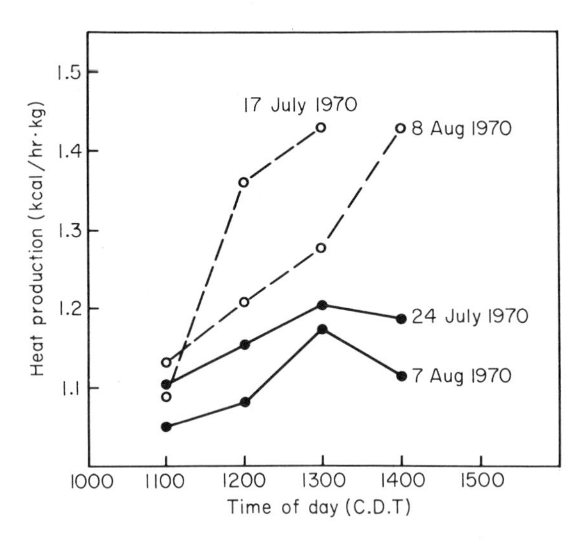

FIG. 3. Heat increment of a high protein meal in subject JS. ○–○, *High protein meal at time 0;* ●–●, *no food eaten. Unpublished data of Bailey* et al. *(20a).*

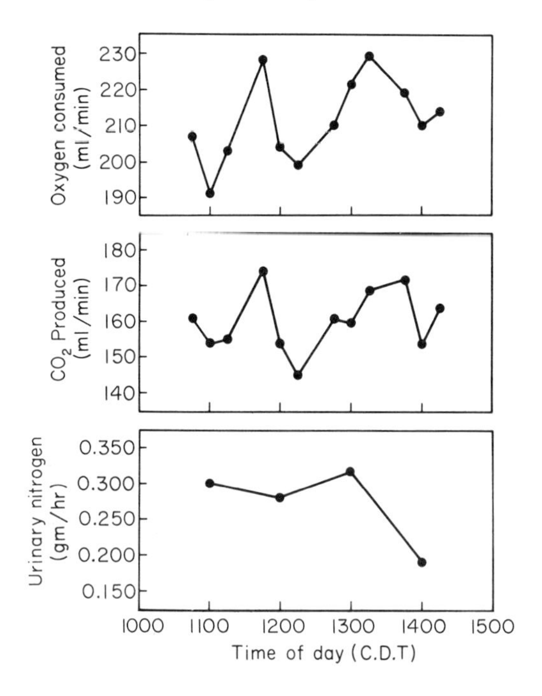

FIG. 4. Cyclical variations in oxygen consumption, carbon dioxide production, and nitrogen excretion during 5 hr after 12 hr in the postabsorptive state. Subject DB. Unpublished data of Bailey et al. *(20a).*

consistently show three things about the SDA of fat. First, at rest it is not easy to demonstrate in all subjects all the time (Fig. 5) (20a), in part because of the variability and cyclical variation in the resting metabolism taken as the base line. Second, and regularly, the SDA of a fat meal disappears during walking. This could mean one of at least three things—suppression of metabolism of fat, suppression of absorption of fat, or suppression of other metabolism by replacement with fat. Regardless of the explanation, the heat increment of a fat meal is not additive to the walk. Third, and regularly, the heat increment of the fat meal is accentuated in recovery after the walk. The walk per se triggers the SDA of the fat meal. This is reminiscent of the post-exercise ketosis we have studied; it is suppressed in exercise, appearing only in recovery after prolonged exercise.

We (16) and others (22) have speculated that prolonged moderate exercise triggers the endocrine system somatotrophin-insulin-glucagon so that free fatty acids are mobilized and used by the working muscle. These acids continue to circulate after the exercise, and ketosis ensues. Apparently so does the heat increment of a fat meal.

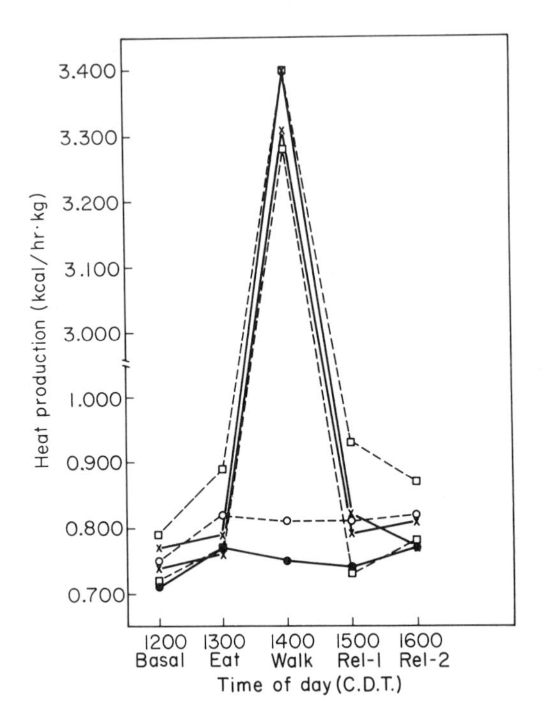

FIG. 5. Experiments on subject DB. Each lasted 5 hr, hour 1 always being a postabsorptive rest and hours 4 and 5 being rest in recovery. Hour 2 was "no eat" or "eat" a high-fat meal. Hour 3 was "no walk" or "walk" at 5 km/hr on a level treadmill at 20°C and RH 50%. •, No eat-no walk; ○, eat-no walk; x no eat-walk; □, eat-walk. – – –, eating; —, no eating.

Sargent and Johnson (1, 2) have argued that in general the SDA of any nutrient by virtue of its being nonshivering thermogenesis might be of importance in releasing calories that could be used by the muscles. For fat we are postulating a direct relation between the exercise and the heat increment. At present we are beginning to explore the relationship between fat metabolism, work and environment. We know that exposure to cold per se (18) is ketogenic, and with respect to post-exercise ketosis, exposure to heat is antiketogenic. We feel that there is bound to be an intimate relation between fat metabolism in general, prolonged muscular work, and metabolism.

Conclusion

For separate reasons, I have chosen to discuss two aspects of nutrition–metabolism and exposure to heat. The first, fulfilling water requirements, is the overpowering primary nutritional problem in the heat. The second, the heat increment of fat, represents an obscure area in our knowledge of the interplay among nutrition, metabolism, work, and the thermal environment. Much important research still lies ahead, I believe, especially in the area of neurohumoral controls of metabolism and thermoregulation.

References

1. Sargent, F., and Johnson, R. E. (1957). "The Physiological Basis for Various Constituents in Survival Rations," Part IV. An Integrative Study of the All-Purpose Survival Ration for Temperate, Cold, and Hot Weather. WADCTR 53-484. Wright-Patterson, AFB, Ohio.
2. Johnson, R. E., and Sargent, F., II. (1958). Some quantitative interrelationships among thermal environment, human metabolism and nutrition. *Proc. Nutr. Soc.* **17**, 179.
3. Pitts, G. C., Johnson, R. E., and Consolazio, F. C. (1944). Work in the heat as affected by intake of water, salt and glucose. *Amer. J. Physiol.* **142**, 253.
4. Adolph, E. F. *et al.* (1947). "Physiology of Man in the Desert." Wiley (Interscience), New York.
5. Gamble, J. L. (1954). "Chemical Anatomy of Extracellular Fluid," 6th ed. Harvard Univ. Press, Cambridge, Massachusetts.
6. Dill, D. B. (1938). "Life, Heat and Altitude." Harvard Univ. Press, Cambridge, Massachusetts.
7. Johnson, R. E., Wakat, D., Gerber, L., and Woodall, T. (1971). Water economy in long distance running. *Proc. Int. Congr. Regul. Food Water Intake,* in press.
8. Sanctorius, S. (1614). "De Statica Medicina." Venice, Italy.
9. Peters, J. P. (1935). "Body Water. The Exchange of Fluids in Man." Thomas, Springfield, Illinois.
10. Passmore, R., Meiklejohn, A. P., Dewar, A. D., and Thow, R. K. (1955). An analysis of the gain in weight of overfed thin young men. *Brit. J. Nutr.* **9**, 27.
11. Consolazio, C. F., Johnson, R. E., and Pecora, L. J. (1963). "Physiological Measurements of Metabolic Functions in Man." McGraw-Hill, New York.
12. Pugh, L. G. C. E., Corbett, J. L., and Johnson, R. H. (1967). Rectal temperatures, weight losses, and sweat rates in marathon running. *J. Appl. Physiol.* **23**, 347.
13. Mitchell, H. H. (1963, 1964). "Comparative Nutrition of Man and Domestic Animals," Vols. 1 and 2. Academic Press, New York.
14. Mitchell, H. H., Glickman, E. H., Lambert, H., Keeton, R. W., and Fahnestock, M. K. (1946). The tolerance of man to cold as affected by dietary modification: Carbohydrate versus fat and the frequency of meals. *Amer. J. Physiol.* **148**, 84.
15. Rochelle, R. H., and Horvath, S. M. (1969). Metabolic responses to food and acute cold stress. *J. Appl. Physiol.* **27**, 710.
16. Passmore, R., and Johnson, R. E. (1958). The modification of post-exercise ketosis (the Courtice-Douglas effect) by environmental temperature and water balance. *Quart. J. Exp. Physiol.* **43**, 352.
17. Courtice, F. C., and Douglas, C. G. (1936). The effect of prolonged muscular exercise on the metabolism. *Proc. Roy. Soc., Ser. B* **119**, 381.
18. Johnson, R. E., Passmore, R., and Sargent, F., II. (1961). Multiple factors in experimental ketosis. *Arch. Intern. Med.* **107**, 43.

19. Kart, R. M., Johnson, R. E., and Lewis, J. S. (1945). Defects of pemmican as an emergency ration for infantry troops. *War Med.* **7**, 345.
20. Johnson, R. E., Robbins, F., Schilke, R., Molé, P., Harris, J., and Wakat, D. (1967). A versatile system for measuring oxygen consumption in man. *J. Appl. Physiol.* **22**, 377.
20a. Bailey, D. W., Harry, D., Johnson, R. E., and Kupprat, I. C. (1971). Unpublished data.
21. Buskirk, E. R., Iampietro, P. F., and Welch, B. E. (1957). Variations in resting metabolism with changes in food, exercise and climate. *Metab. Clin. Exp.* **6**, 144.
22. Johnson, R. H., Walton, J. L., Krebs, H. A., and Williamson, D. H. (1969). Post-exercise ketosis. *Lancet* **2**, 1383.

VIII

Large Mammals in the Desert

D. H. K. LEE

Introduction

In summarizing the work of his own group in his 1938 monograph (1), Dill casts Buxton as the forerunner of studies on animal life in deserts, with the publication in 1923 of a book under that title (2), prefaced with the statement that the "study of desert creatures and their environment . . . can only be approached with the help of botanists, meteorologists, physicists, and others."

From Buxton, through Dill, that change has continued through several of those who had the fortune to work with the Fatigue Laboratory team, either in the laboratory or in one of the numerous, demanding, and eminently successful field forays that emanated therefrom. My own desert studies have been almost entirely confined to man (4)–not entirely a typical mammal! Nevertheless, I believe that certain principles established in those studies are also applicable to large mammals in general. In conjunction with experience from animal studies in warm humid conditions, and a general interest in comparative vertebrate physiology–long ago urged upon me by Dill–they point up important facets of their desert problems. Inevitably, however, I shall draw extensively from the reports of Schmidt-Nielsen and his colleagues.

Unlike many small animals who can hide in shade, burrow in the ground, or seek out damp spots, large mammals must face the full force of desert conditions, with high intensities of solar radiation, strong winds, highly reflective terrain, clear skies, high air temperatures, and low humidities. Moreover, having adapted to these conditions, as the survivors must, they also face sharply reversed conditions at night, with strong radiation gradients to the sky and markedly lower air temperatures. It is a tribute to the plasticity, as well as the adaptability, of mammalian inheritance that so many forms have succeeded, each in its own way. Four aspects of the action-reaction complex will be examined.

Exposure to Solar Radiation

The functionally quadruped and the functionally biped mammals (not forgetting the kangaroo) expose very different profiles to the summer sun. This can be demonstrated by considering the total incidence of solar radiation on a cylinder of given dimensions, first in the vertical position and then in the horizontal position. In the case of the horizontal cylinder, orientation to the sun is very important. A cylinder that moves in the horizontal plane so that the end is always pointed toward the sun will have a minimum exposure, whereas one whose side always faces the sun will have a maximum. A randomly oriented cylinder will be exposed to an intermediate incidence. The relevant calculations are as follows.

The incidence of solar radiation on an object can be computed as the sum of that which falls on a horizontal and that which falls on a vertical projection of the object, respectively.

For a *vertical* cylinder the total incidence is given by

$$R_{ivc} = R_{sh}(\pi r^2 + 2rl \cot h) \tag{1}$$

For a *horizontal* cylinder equioriented with respect to the sun

$$R_{ihc} = R_{sh}\left[2rl + \left(\frac{4rl}{\pi} + 2r^2\right)\cot h\right] \tag{2}$$

where R_{sh} is incidence on a horizontal place, r is radius, l is length of the cylinder, and h is the solar altitude (angle with horizontal).

A cylinder equivalent in size to man would have $r = 0.15$ m and $l = 1.8$ m. The equations then reduce to

$$R_{ivc} = R_{sh}(0.07 + 0.54 \cot h) \tag{3}$$

for a *vertical* cylinder and

$$R_{ihc} = R_{sh}(0.54 + 0.39 \cot h) \tag{4}$$

for a *horizontal* cylinder.

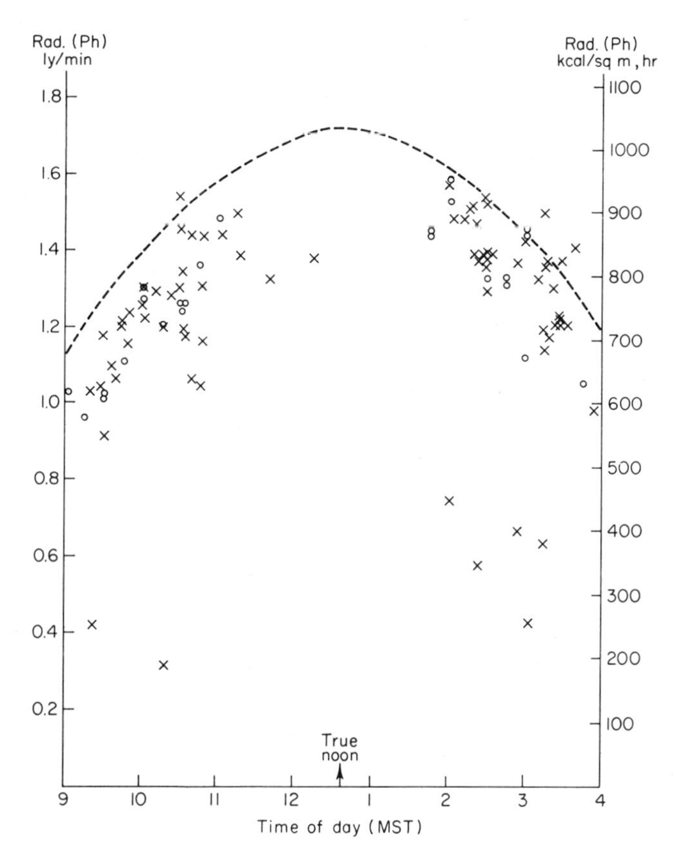

FIG. 1. Comparison between maximum intensity of solar radiation as calculated by Klein's equations (3) and observations made during two summers (1952, 1953): ○, *1952;* ×, *1953;* - - -, *calculated maximum. [Reproduced from Lee (4) by permission of American Physiological Society.]*

It will be noticed that for a vertical cylinder there is a small constant term representing the horizontal projection and a large coefficient for the term varying inversely with the solar altitude. For the horizontal cylinder the constant term is large and the coefficient for the variable term smaller.

The calculated maximum and observed values for R_{sh} at Yuma, Arizona, in midsummer are given in Fig. 1.

The solar altitude (h) is given by the equation

$$\sin h = \sin d \sin l + \cos d \cos l \cos t \tag{5}$$

where d is declination of the sun for the season, l is latitude of the place, and t is hour angle (angular displacement of the sun from the meridian).

For Yuma in midsummer this reduces to

$$\sin h = 0.21 + 0.78 \cos t \tag{6}$$

Using the calculated maximum values of R_{sh} from Fig. 1 and the values of h given by Eq. (6), the incidence of solar radiation on vertical and equioriented horizontal cylinders, between the hours of 9 a.m. and 4 p.m., in Yuma at midsummer, is as shown in Fig. 2.

The contrast in solar radiation on the horizontal and vertical stances is immediately apparent. The quadruped is exposed to a high incidence throughout the day, with little decline at noon. The incidence on the biped is at all times lower than on the quadruped and undergoes a marked reduction in the middle of the day. Both may benefit from the warming effect in the earlier hours, but both are likely to be greatly stressed in the midafternoon hours when air temperature and ground temperature are also high and adding their quota to the total heat load, as shown in Table I.

The importance of shade, particularly in the afternoon, is readily deduced. Natural shade, however, is not very common under typical desert conditions. When artificial shade is provided by man, it is important to see that the shading device does not itself become a heat trap. Walls set up to screen out reflected

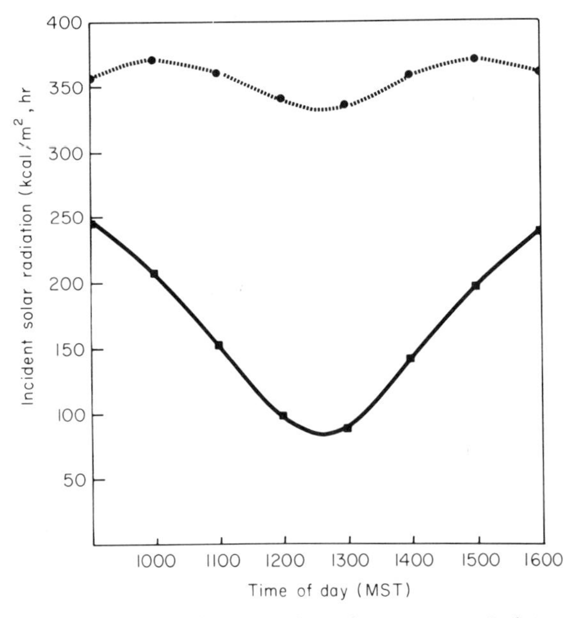

FIG. 2. Comparison of incidence of direct solar radiation on vertical (upper curve) and equioriented horizontal cylinders (lower curve) of human dimensions, calculated from maximum intensity given by Klein's equations (3) for Yuma, Arizona, in July.

Table I. Determination of Relative Heat Strain (RHS) in Hypothetical Cows and Man for Midsummer Conditions at Yuma, Arizona

	Time of day (MST)							
	0900	*1000*	*1100*	*1200*	*1300*	*1400*	*1500*	*1600*
Air temp. °C	*32.0*	*34.5*	*36.8*	*38.0*	*39.0*	*39.5*	*39.5*	*39.0*
Quadruped[a]								
Radiation incid. (kcal/m², hr)[a]	356	370	360	340	335	360	370	360
Equivalent increm. to air temp. (°C)	13.0	13.5	13.0	12.0	12.0	13.0	13.5	13.0
Air temp. entry to RHS chart (°C)	45.0	48.0	49.8	50.0	51.0	52.5	53.0	52.0
RHS (low-sweating cow)	1.27	1.38	1.46	1.47	1.50	1.57	1.60	1.54
RHS (high-sweating cow)	0.78	0.85	0.89	0.90	0.91	0.95	0.96	0.93
Biped								
Radiation incid. (kcal/m², hr)[a]	245	205	150	100	90	140	195	2.40
Equivalent increm. to air temp. (°C)	7.25	5.25	2.50	0	−0.50	2.00	4.75	7.00
Air temp. entry to RHS chart (°C)	39.25	39.75	39.30	38.00	38.50	41.50	44.25	46.00
RHS (man)	0.65	0.67	0.65	0.62	0.63	0.71	0.76	0.81

[a]To convert kcal/hr to watts, multiply by 1.163.

solar and longwave radiation from hot ground may easily do more harm than good by trapping heat absorbed by the roof (4).

Thermal Stress and Strain

Following the concept introduced by Belding and Hatch (5), and elaborated by Henschel and myself (6), I have used a measure termed "Relative Strain" to express the disturbance likely to be experienced by an animal exposed to hot conditions (7).

Relative Strain is the ratio between the evaporative cooling required to maintain thermal balance and the maximum evaporative cooling that the animal can enjoy.

$$\text{Relative Strain} = \frac{\text{evaporative cooling required to maintain thermal balance}}{\text{maximum evaporative cooling available to the animal}} \quad (7)$$

From the well-known equations for heat transfer between an animal and its environment (8), the components of this ratio can be written.
Evaporative cooling required

$$E_{req} = \left\{ A_e \left[h_{cs}(T_a - T_s) + RI_a \right] / (I_a + I_{cw}) \right\} + M \quad (8)$$

Maximum physiological evaporative cooling possible

$$E_{max} = A_e(h_{es}/ra) \cdot (P_{ss} - P_a)w \cdot [I_a/(I_a + I_{cs})] + h_{er} \cdot \dot{V} \cdot (P_{bs} - P_a) \quad (9)$$

where T_a is air temperature; T_s is skin temperature; R is net rate of radiation gain; I_a is resistance to conduction offered by the air; I_{cw} is that offered by the wet coat (or clothing); A_e is surface area available for evaporation; P_{ss} *is* saturation vapor pressure at skin temperature; P_a is vapor pressure of the air; w is degree of skin wetness; r_a is resistance to water vapor diffusion offered by the air; $\dot{V}$ is respiratory volume; P_{bs} is saturation vapor pressure at body temperature; and h_{cs}, h_{es}, and h_{er} are the heat transfer constants involved.

The first term of the equation for maximum evaporative cooling when $w = 1$ is subject to the constraint that its value cannot exceed that obtainable from the maximum rate at which the animal can provide water for evaporation from the skin. This varies with the density, secretory capacity, and physical nature of the secretion provided by sweat glands. All mammals other than rodents and lagomorphs have sweat glands, but their ability to provide fluid for evaporation varies very greatly with the species and even with strains within species (9).

If the metabolic rate, surface area, coat characters, maximum respiratory volume, maximum sweat rate, air movement, and radiation load are known, then the relative strain can be expressed simply in terms of the air temperature and vapor pressure to which the animal is exposed. The combinations of air temperature and vapor pressure that will produce a given degree of relative strain can be shown on a psychrometric chart as an angulated line, the upper part relating to evaporative demands within the animal's sweating capacity and the lower to demands in excess of that. Figures 3 and 4 show such lines for various degrees of relative strain for two hypothetical cows, which differ only in maximum sweating capacity.

The advantage of the higher sweating capacity is easily seen. For example, an air temperature of 35°C and vapor pressure of 15 mm Hg results in a relative strain of 0.91 in the first animal, but only 0.56 in the second. The first is close to an intolerable limit, the second well within the physiological range. The comparable condition in a roughly equivalent man (Fig. 5) is similar to that for the second cow, viz., 0.51.

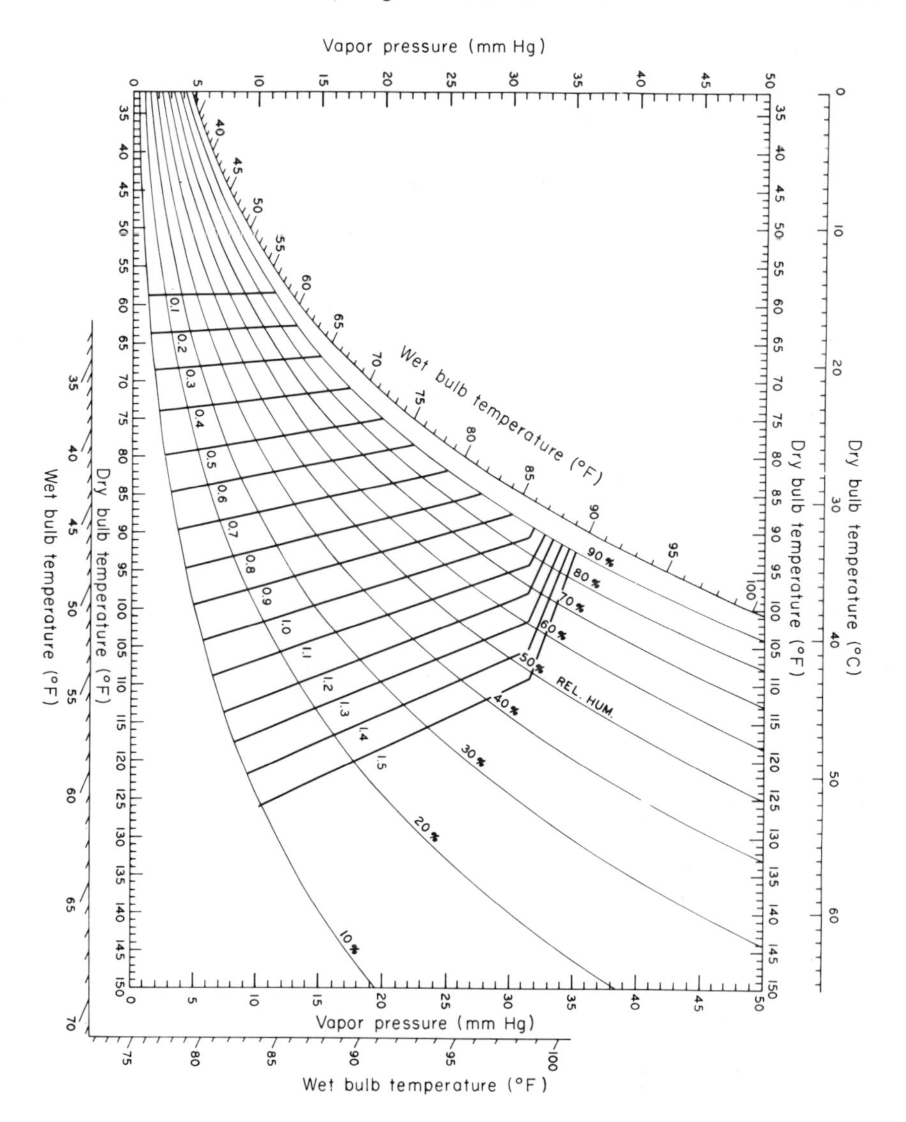

FIG. 3. Psychrometric chart for estimating the relative heat strain that would be developed in a cow of surface area 4.5 m² exposed to various combinations of air temperature and humidity. Metabolic rate 100 kcal/m² · hr; air movement 45 m/min; maximum sweating capacity (S_{max}) 222 gm/m² · hr. $M = 100$; $I_{cw} = 0.3$; $V_{max} = 16{,}000$; $A = 4.5$; $I_a = 0.5$. [Reproduced from Lee (7) by permission of International Society of Biometeorology.]

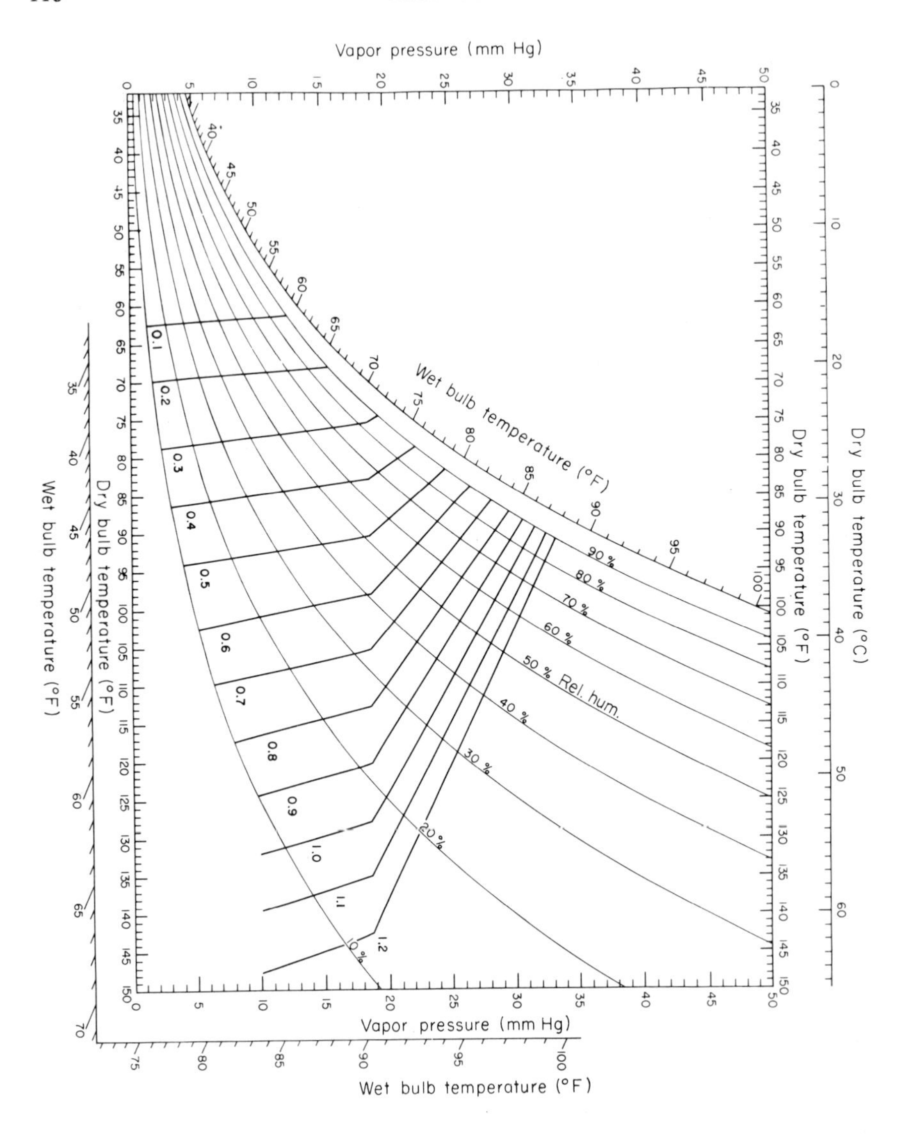

FIG. 4. Similar chart for cow with maximum sweating capacity (S_{max} = 444 gm/m² · hr). [Reproduced from Lee (7) by permission of International Society of Biometeorology.]

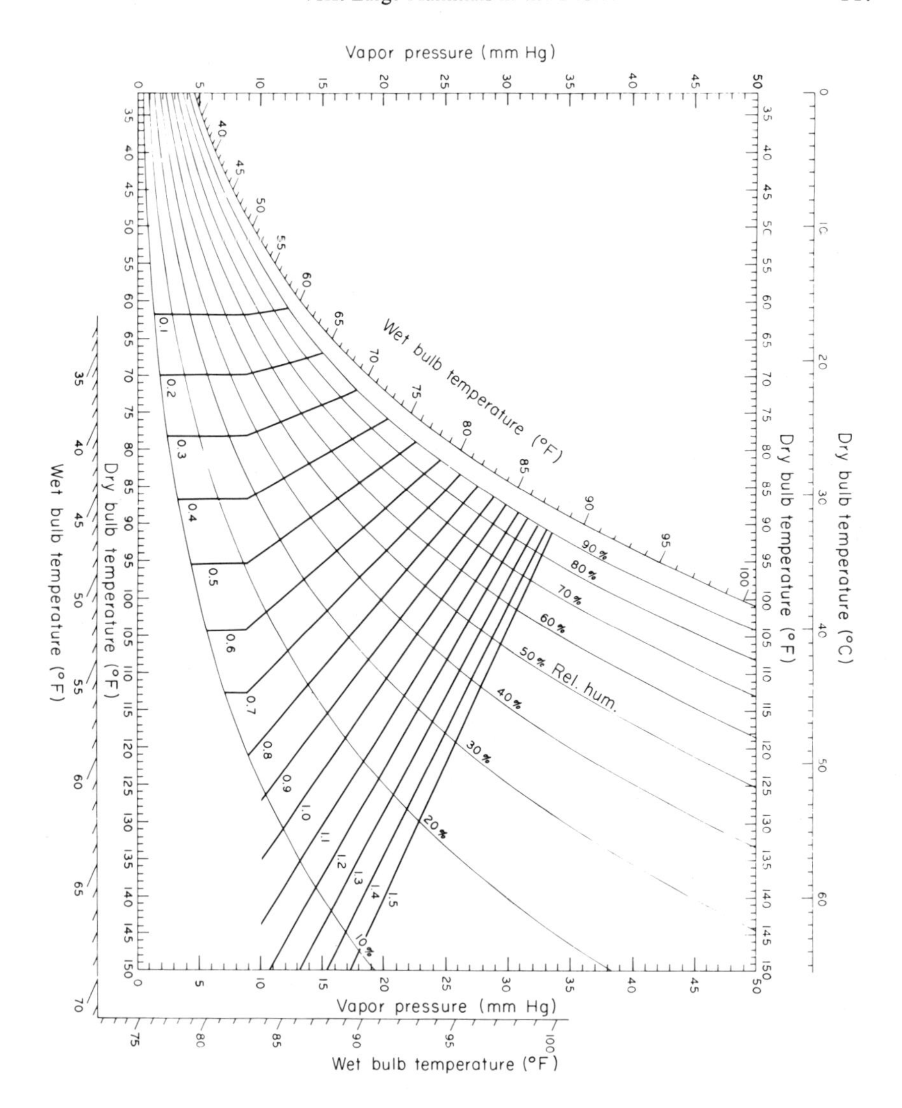

FIG. 5. Similar chart for man of smaller size (surface area 1.75 m^2) and maximum sweating capacity (S_{max}) of 570 gm/m^2 · hr. [Reproduced from Lee (7) by permission of International Society of Biometeorology.]

The effect of radiant heat loads greater than that used in constructing the chart can be estimated by calculating from Eq. (8) the rise in temperature that would produce the same increase in E_{req}. For the air movement postulated in developing Figs. 3-5, and assuming a 33% reflection of incident radiation, the equivalent rise of air temperature would be about 1°C for every 20 kcal/m^2 · hr. That would be equivalent to raising the prevailing air temperature for quadrupeds by 12.5°C for most of the day, but only by 7.5°C for bipeds in midafternoon, and not at all at midday. It is interesting to note that measurements of increment of air temperature equivalent to radiation load (ITER) at Yuma, using human physiological responses as the criterion of equivalence, ranged from 6.8° to 9.8°C, depending upon the clothing and air movement used (10).

The conditions prevailing at Yuma in the course of a midsummer day, the resultant equivalent air temperature taking into account the radiation load, and the relative heat strain index numbers for the three mammals are given in Table I and Fig. 6.

The low-sweating cow would be past its powers of compensation by evaporative cooling well before the "working" day began at 9 a.m. The high-sweating cow, by contrast, would be approaching its compensatory limit only at the hottest hour of 3 p.m., even though its ability to supply sweat is still inadequate to the full task. Man, by virtue of his upright stance, plus his ability to sweat heavily, is quite capable of handling the situation, although at a price, particularly in the midafternoon.

Evaporative Cooling

The importance of evaporative cooling in tolerance for hot desert conditions is amply illustrated by the example just given. The amount of water available by transudation through skin from deeper tissues without the agency of sweat glands (true perspiration) is quite small—about 30 gm/m^2 · hr, providing only 15 kcal/hr of cooling. The amount available for evaporation into inspired air is limited by the animal's mechanical abilities for increased respiration, the increased metabolic heating resulting from the effort of hyperventilation, and the risk of alkalosis. The last limitation is reduced if the depth of respiration is reduced as the rate is increased. The ideal relationship is given by

$$R = (V - A)/d \tag{10a}$$

or

$$t = d \cdot V/(V - A) \tag{10b}$$

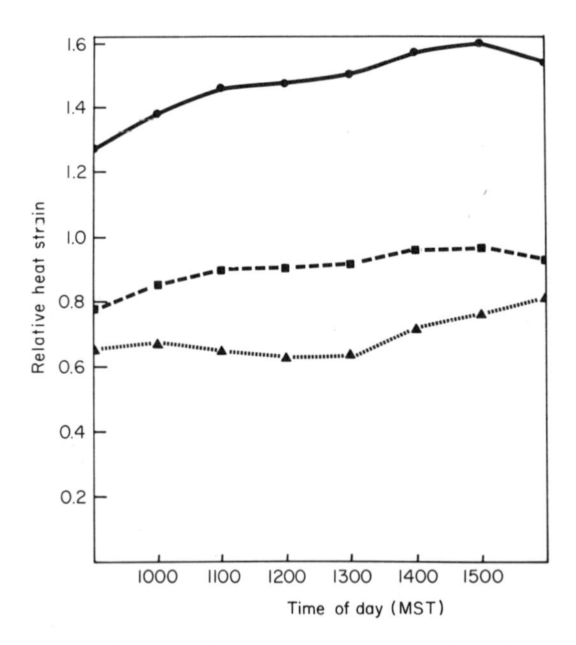

FIG. 6. Relative heat strain experienced by the two cows and man postulated in previous figures under radiative and temperature conditions prevailing over the working day in Yuma, Arizona, in July. Low-sweating cow, upper curve; high-sweating cow, middle curve; man, lower curve.

where R is required respiratory rate, V is desired respiratory minute volume, A is normal alveolar ventilation rate, d is volume of dead space, and t is tidal volume.

To meet this ideal condition the respiratory rate would have to increase much more rapidly than the ventilation volume. Given a dead space of 0.15 liter, and a normal alveolar ventilation of 7 liters/min, the required respiratory rate would rise 11 times, from 20 to 220, as the respiratory minute volume is quadrupled from 10 to 40 liters/min.

The inertia of the respiratory organs is such that the high respiratory rate necessary to maintain normal alveolar ventilation as the minute volume is increased can seldom be achieved, although the rabbit comes close at several hundred respirations per minute. The restriction is removed, however, if the normal dead space can be greatly increased and if there is assurance of a sufficient supply of water for evaporation, as in the dog in panting (11). Animals that do not pant but inspire through the nose and exhale through the mouth still retain some advantage by cooling in the nasal passages on inspiration and in the buccal cavity on expiration (12). In general, however, the respiratory mode of evaporative cooling has distinct limitations.

Glands that can deliver water to the surface of the skin when and as required, without the mechanical inconveniences and drawbacks of strong respiratory effort, provide a much more efficient mode of evaporative cooling. As pointed out by Bligh (9), mammalian epitrichial glands range in function from truly apocrine-type excretion of moisture-poor cell content to eccrine-type excretion of fluid, with a corresponding range of thermolytic usefulness. The advantage is increased with the human atrichial-type gland, which delivers its eccrine secretion directly to the skin surface without hindrance by follicles and sebaceous secretion.

Let it suffice to say that even the most efficient mechanisms have their limitations, without attempting to review the extensive body of work that has been done on the detailed physiology of sweat glands and sweating (13-16). But it is very suggestive that the best mechanisms are just good enough to get their possessors through the rigors of naturally occurring conditions. The selection pressure of those natural conditions seems to have played a determinant role; a greater capability would have been required so seldom that there would have been little drive for its emergence.

A subsidiary physiological mode of providing evaporative cooling is by the spreading of saliva on the fur. This is frequently seen in cats and the smaller marsupials. This is not often seen in the pig, which is among the most prolific of salivators (17).

For every adaptation there is a price to be paid. For evaporative cooling it is the replenishment of the water that is lost, and the consequences of any deficit. To this, in the case of sweating man, is added a threat of possible salt deficiency. In most animals the initial water deficit is taken by the circulating blood volume, so that circulatory insufficiency easily results. In man, gravitational forces then quickly produce disturbances of cerebral function. Schmidt-Nielsen has shown, however, that in the camel the blood volume is retained at normal levels and the dehydration passed on to the cells (18, 19). It would be interesting to know what happens in the giraffe! To the extent that dehydration develops in a tissue, virtually all of its functions are likely to be affected. Reduction of urine volume, an early and marked effect, may threaten elimination of metabolites, but this also is minimized in the camel through recycling of retained urea by the ruminant stomach into amino acids and protein (20). The consequences of dehydration, as seen in man, are given in Table II (21).

The production of metabolic water as a result of oxidation, particularly of fats, is often cited as countering dehydration, but Schmidt-Nielsen points out (19) that the increased respiration required by the additional supply of oxygen, and the additional heat production, would lead to a still greater evaporation. The mysterious ability of some animals to live in the desert with little or no drinking water available usually turns out to be due to a saving in water use rather than to an access to internal sources of water; this is usually at the expense of some other function, such as control of body temperature. Camels do not pant (18).

Table II. Symptoms Characteristic of Desert Dehydration in Man (21)

Water deficit (% body wt.)	*Symptoms*
1-5	Thirst Discomfort and complaints Anorexia Apathy
5-10	Stumbling Headache Indistinct speech Dyspnoea Cyanosis
10-15	Spasticity Delirium Inability to swallow Shriveled skin Sunken eyes
>15	Deafness Numb skin Stiffened eyelids Cracked skin Anuria

Dehydration may raise the threshold body temperature at which sweating or panting is initiated (19). It may also reduce food intake and thus the metabolic substrate, heat production, and respiratory requirements (19). Animals that can accept dehydration often show remarkable powers of rapid rehydration when water is available (18-20).

Other Modes of Adaptation

To the extent that evaporative cooling is not available or if its use would bring about undesirable effects, recourse must be had to other modes of adaptation if hot desert conditions are to be withstood.

As pointed out earlier, large mammals have little opportunity for escape from open desert conditions into sheltered niches. Quadrupeds could, of course, reduce insolation by continuously orienting their narrow end to the sun. This has been reported for the camel (19), but does not seem to be a characteristic behavior of horses, cattle, or sheep. Postural reductions of exposed surface area also seem to be peculiar to the camel. Metabolic heat production can be reduced by lowered activity, relaxed posture, or endocrine control of resting metabolic rate. These possibilities seem not to have been examined to any great extent. One would think that biting flies and other pests would often make relaxation difficult.

Insulation can provide a more effective means of reducing the heat burden. With a high incident radiation load, and air temperature greater than skin temperature, both terms in the numerator in Eq. (8) are positive, so that insulation provided by the coat will reduce heat gain. Any offset by resistance to water vapor loss in Eq. (9) is minimized by the large differential in vapor pressure between skin and air. Various experiments have shown the beneficial effects of a heavy wool coat for the heat balance of sheep under hot dry conditions (Fig. 7) (22). The insulating coat, however, must be compact. Loose

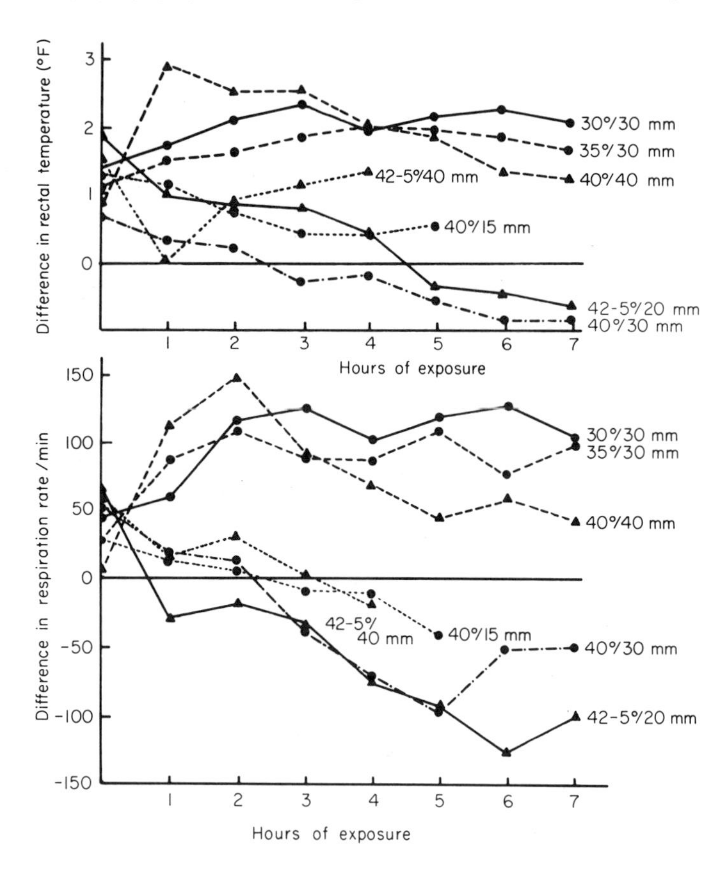

FIG. 7. Difference between reactions (rectal temperature and respiratory rate) of unshorn and shorn sheep exposed to hot conditions. Unshorn animals often show less severe reactions at higher air temperatures, particularly when the humidity is low. [Reproduced from Lee (22) by permission of CSIRO.]

fibers will be disturbed by the wind, with loss of insulative value, and penetrating solar radiation will be absorbed close to the skin. The color of the coat probably makes little difference to the heat absorption. Dark wool, for instance, absorbs more of the incident radiation, but near the surface whence it can be removed by the normally strong air movement. Even with thin materials, measurements show that transmission of solar heating to underlying "skin" is only slightly greater with black than with white fabrics of identical weave (23).

The smaller ratio of surface area to body mass in large animals can work to their advantage under desert conditions, decreasing heat gain by conduction from hot air and by radiation from heated terrain. Similarly, appendages are less desirable here than they are under warm humid conditions.

The significance of a rising body temperature is somewhat reduced if the most susceptible or most critical internal organs can be spared. A remarkable provision for sparing the brain has been described for the donkey, the sheep, the goat, the ox, and the pig [Fig. 8 (24, 25)]. In these animals, arterial blood brought to the cranium through the internal maxillary artery flows in countercurrent fashion beside cooled venous blood from the extensive nasal passages, in the rete of the cavernous sinus, before it goes to the brain. Evaporative cooling from nasal inspiration is thus directed largely to the benefit of the brain.

Finally, use is made by some animals, such as the oryx and eland, of the diurnal variation in ambient conditions. By allowing the core temperature to fall by night, a relatively large animal can use heat storage to take up a large part of the daytime load (18, 25). If it can also protect vital organs, it can go further and let the core temperature rise to relatively high levels (40°C) in midafternoon. Smaller animals, such as the gazelle, have less margin to play with.

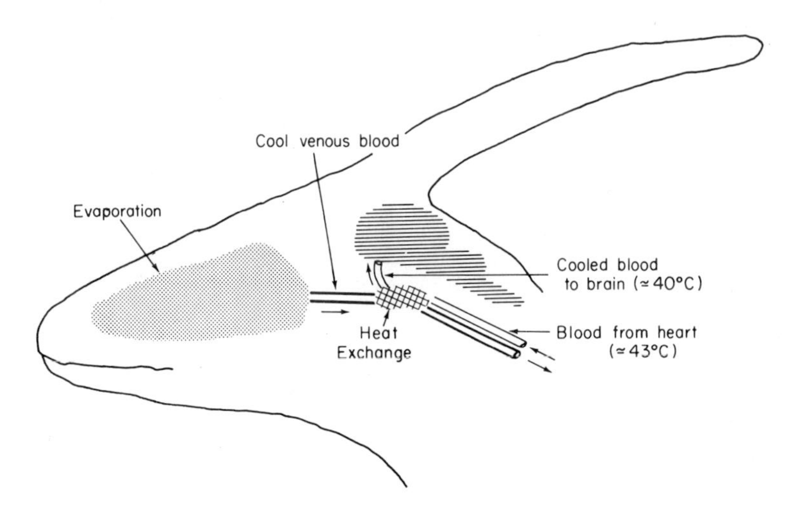

FIG. 8. Countercurrent cooling of arterial blood to brain by venous return from nasal cavities in the oryx. (Personal courtesy of C. R. Taylor.)

Conclusion

The incidence of solar radiation in the hot desert is least on the biped, particularly at noon, and greatest on the quadruped, which exposes its flank to the sun. An equioriented quadruped receives about the same radiant load throughout the major part of the day. The heat stress imposed on an animal by the high solar load and high air temperatures is such that considerable evaporative cooling is required to maintain a balance. The evaporative cooling that can be provided from the respiratory tract is limited and likely to bring alkalosis in its train. Cooling by evaporation of water from the skin avoids this difficulty, but the supply of water from sweat glands is limited. If the water loss is not replenished, the resultant dehydration may bring about some savings in water loss, but at the expense of a rising body temperature. Many other adverse effects of dehydration may supervene. Supplementary ways in which crises may be avoided include using the heat storage capacity of the body to offset daytime heating with nighttime cooling; transfer of heat from arterial blood to cooled nasal blood before it enters the brain; selective maintenance of blood volume in spite of dehydration; maintenance of pelage insulation, orientation of minimum profile to the sun, and utilization of such shade as can be found. The particular combination of defensive and compensatory mechanisms used by animals that manage to survive under desert conditions varies greatly. There is room for a much more systematic study of these patterns.

References

1. Dill, D. B. (1938). "Life, Heat and Altitude." Harvard Univ. Press, Cambridge, Massachusetts.
2. Buxton, P. A. (1923). "Animal Life in Deserts." Arnold, London.
3. Klein, W. H. (1948). Calculation of solar radiation intensity and the solar heat load on man at the earth's surface and aloft. *J. Meteorol.* **5**, 119.
4. Lee, D. H. K. (1964). Terrestrial animals in dry heat: Man in the desert. *In* "Handbook of Physiology" (Amer. Physiol. Soc., D. B. Dill *et al.* eds.), Sect. 4, p. 551. Williams & Wilkins, Baltimore, Maryland.
5. Belding, H. S., and Hatch, T. F. (1955). Index for evaluating heat stress in terms of resulting physiological strains. *Heat., Piping, Air Cond.* **27**, 129.
6. Lee, D. H. K., and Henschel, A. (1966). Effect of physiological and clinical factors on response to heat. *Ann. N. Y. Acad. Sci.* **134**, 743.
7. Lee, D. H. K. (1965). Climatic stress indices for domestic animals. *Int. J. Biometeorol.* **9**, 29.
8. Lee, D. H. K. (1968). Principles of homeothermic adaptation. *In* "Biometeorology" (W. P. Lowry, ed.), p. 133. Oregon State Univ. Press, Corvallis.
9. Bligh, J. (1967). A thesis concerning the processes of secretion and discharge of sweat. *Environ. Res.* **1**, 28.
10. Lee, D. H. K., and Vaughan, J. A. (1964). Temperature equivalent of solar radiation on man. *Int. J. Biometeorol.* **8**, 61.

11. Dill, D. B., Bock, A. V., and Edwards, H. T. (1933). Mechanisms for dissipating heat in man and dog. *Amer. J. Physiol.* **104**, 36.
12. Schmidt-Nielsen, K., Bretz, W. L., and Taylor, C. R. (1970). Panting in dogs: Unidirectional flow over evaporative surfaces. *Science* **169**, 1102.
13. Robinson, S. (1962). The regulation of sweating in exercise. *Advan. Biol. Skin* **3**, 152.
14. Robinson, S., and Robinson, A. H. (1954). Chemical composition of sweat. *Physiol. Rev.* **34**,202.
15. Dobson, R. L. (1962). The correlation of structure and function in the human eccrine sweat gland. *Advan. Biol. Skin* **3**, 54.
16. Dobson, R. L. (1967). Sweat sodium secretion in normal men, women, and children. *Bibl. Paediat.* **86**, 23.
17. Robinson, K., and Lee, D. H. K. (1941). Reactions of the pig to hot atmospheres. *Proc. Roy. Soc. Queensland* **53**, 145.
18. Schmidt-Nielsen, K. (1959). The physiology of the camel. *Sci. Amer.* **210**, 140.
19. Schmidt-Nielsen, K. (1962). Comparative physiology of desert mammals. *Mo., Agr. Exp. Sta., Spec. Rep.* **21**.
20. Schmidt-Nielsen, K. (1956). Animals and arid conditions: Physiological aspects of productivity and management. *In* "Future of Arid Lands," Publ. No. 43, p. 368. Amer. Ass. Advan. Sci., Washington, D.C.
21. Adolph, E. F. (1947). "Physiology of Man in the Desert." Wiley (Interscience), New York.
22. Lee, D. H. K. (1950). Studies of heat regulation in the sheep, with special reference to the merino. *Aust. J. Agr. Res.* **1**, 200.
23. Breckenridge, J. R., Pratt, R. L., and Woodcock, A. H. (1960). Effect of clothing color on heat load from solar radiation. *Fed. Proc., Fed. Amer. Soc. Exp. Biol.* **19**, 178.
24. Baker, M. A., and Hayward, J. N. (1968). The influence of the nasal mucosa and the carotid rete upon hypothalamic temperature in sheep. *J. Physiol. (London)* **198**, 561.
25. Taylor, C. R. (1969). The eland and the oryx. *Sci. Amer.* **220**, 88.

IX

Small Mammals in the Desert

W. G. BRADLEY and M. K. YOUSEF

Introduction

Deserts are highly demanding ecosystems in which biota must cope with or avoid a variety of environmental problems–heat and cold, scarcity of water, and a highly variable food supply. Yet deserts are characterized by a varied and relatively rich, small mammal fauna. The physiological adaptations of small mammals in the desert have been an area of continuing research since the pioneering work of the Schmidt-Nielsens on the kangaroo rat (1). Recent reviews discussing physiological adaptation in desert mammals tend to reduce the value of another general review at this time (2–5). We have, therefore, tried to restrict our coverage to areas that have been adequately reviewed, in an effort to emphasize some ecological, behavioral, and physiological aspects of adaptation. In order to adequately discuss small mammals in the desert we have restricted our discussion to bats and rodents of North American deserts; both show a variety of adaptive features that allow them to successfully evade, circumvent, or cope with desert environments.

Faunal Analysis

North American Cenozoic fossil floras have been used to evaluate past climates and environments (6). Early in the Cenozoic, tropical environments were widespread with cooler climates indicative of north temperate conditions restricted to high latitudes. During the Eocene and lower Oligocene epochs general cooling resulted in a withdrawal of tropical environments to the south. Continuing continental uplift resulted in cooling, drying, and the development of an arid subtropical area on the Mexican Plateau. Further displacement both northward and southward of boreal and tropical environments was accompanied by expansion of arid subtropical, arid grassland, desert scrub, chaparral, and woodland environments. The four North American deserts appeared as regional units sometime in the Pliocene about 10 million years ago. Further evolutionary refinement has produced the deserts as they are today.

The Great American Desert located in western United States and Mexico is commonly classified as four separate deserts (Fig. 1). Desert environments extend in a continuous north-south gradient from the low, hot Sonoran, to the middle elevational, transitional Mohave, and to the higher, cold Great Basin Desert to the north. The Chihuahuan Desert has a disjunct distribution to the east, and although located in the south, has rather severe winters because of its elevated position.

The distribution of bats and rodents arranged by family is given in Table I. The more southerly located deserts, especially the Sonoran, have far more small mammal species than the more northern deserts. The increased species diversity in southern deserts is related to geographic position, increased habitat diversity, and high level of endemism on Gulf of California islands. Some species are restricted to the southern portion of the Sonoran and, to a lesser extent, the Chihuahuan Desert, having their main distribution in subtropical or tropical

FIG. 1. Location of the deserts of North America. Numbers 1-4 refer to Sonoran, Chihuahuan, Mohave, and Great Basin deserts, respectively.

TABLE I. Distribution of Small Mammals (Bats and Rodents) in North American Deserts

Family	*All deserts*	*Total number of species*			
		Chihuahuan	*Sonoran*	*Mohave*	*Great Basin*
Chiroptera					
Emballonuridae	1	0	1	0	0
Phyllostomidae	7	5	7	1	0
Desmodontidae	1	0	1	0	0
Natalidae	1	1	1	0	0
Mollossidae	5	3	5	3	1
Vespertilionidae	22	15	21	16	13
Total bats	37	24	36	20	14
Rodentia					
Sciuridae	16	5	9	7	5
Geomyidae	5	3	1	1	2
Heteromyidae	35	14	23	10	10
Cricitidae	49	24	34	10	9
Total rodents	105	46	67	28	26

Mexico. Endemics are more limited in their distribution. In contrast, numerous species have a fairly widespread distribution and may be found in all four deserts. It should be emphasized that although the Mohave Desert is small in size, it is transitional between cold and hot deserts, and therefore shares many species with adjacent deserts.

The distribution of faunal units in North American deserts is shown in Fig. 2. Bats inhabiting North American deserts fall into two major groups. Bats with tropical affinities have their main distribution in the southern deserts. Few species reach their northern limits in the Mohave or in the southern edge of the Great Basin Desert. North-temperate bats, however, are the major faunal element in all deserts. Numerous vespertilionid bats (approximately 50%) have their main range in the arid regions of southwestern North America and probably evolved in or adjacent to present-day deserts. The rodent fauna of North America can be divided into three main faunal units (7)—(a) the boreal fauna of high and middle latitudes, (b) the Sonoran fauna of Mexico and western United States, and (c) the tropical fauna of low latitudes and elevations located in Mexico. The boreal rodents make up about 9% of the total rodent fauna. This faunal unit is limited in importance in the southern deserts. The tropical unit makes up about 6% of the rodent fauna and is restricted to the southern deserts. The bulk of the rodent fauna (85%) belongs to the Sonoran faunal unit, which is widespread in North American deserts.

In general, it appears that major centers for the evolution of the small mammal

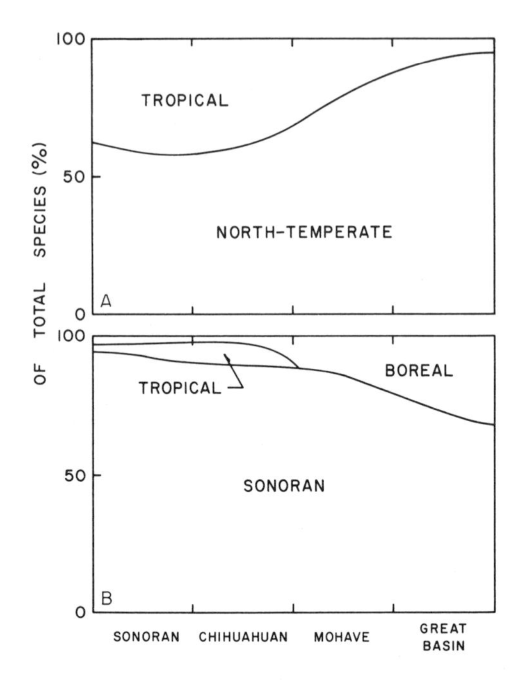

FIG. 2. The distribution of bat (A) and rodent (B) faunal units in the North American deserts.

fauna (bats and rodents) are as follows. (a) The New World tropics, chiefly of Mexico and Central America (approximately 15%), (b) the arid Mexican plateau and southwestern United States (approximately 71%), and (c) north-temperate to boreal regions of North America (approximately 14%). Although tropical affinities in the southern deserts and boreal or north-temperate affinities in the northern desert are both significant, the bulk of the desert fauna has arid subtropical or desert affinities. This characteristic desert fauna exhibits evolutionary adaptations for living under desert conditions. These adaptations may be classified as primarily behavioral or physiological.

Behavioral Adaptations

Whereas the adaptive patterns that allow species to avoid or circumvent the full rigors of desert environments are mainly behavioral, individual functional adjustments usually accompany and complement behavior. Behavioral adaptations used by small desert mammals are varied. Some important behavioral patterns are discussed in the following sections.

ECOLOGICAL DISTRIBUTION AND HABITAT

Desert mammals tend to have ecological distributions along temperature-moisture gradients ranging from hot to mild and arid to mesic. Some species are known to have exacting habitat requirements, but most are widely distributed and occupy various habitats in a region. With few exceptions, most species that evolved in desert or arid regions tend to have less exacting or restrictive habitat requirements. In contrast, most tropical or boreal species have the most exacting habitat requirements in deserts. These habitats tend to be small, disjunct, and more characteristic of and widespread in nondesert regions.

Burrowing is a common characteristic of many small desert mammals. Of the mammals inhabiting forest, steppe, and desert, 6, 47, and 72%, respectively, utilize burrows (8). Boreal and tropical rodents commonly utilize above-ground nests and runways rather than burrows. The size, complexity, and depth of burrows vary greatly depending upon the species and local conditions. Both humidity and temperature are greatly modified in burrows, resulting in a more favorable environment. This is illustrated in Figs. 3 and 4 where yearly temperature patterns for high and low deserts are shown.

Bats commonly utilize various shelters as day roosts and hibernals. Shelters may vary from a tree to a rock crevice, a mine tunnel, or cave. Most shelters in desert areas tend to provide a more uniform and less stressful microenvironment than that characteristic of the surrounding desert.

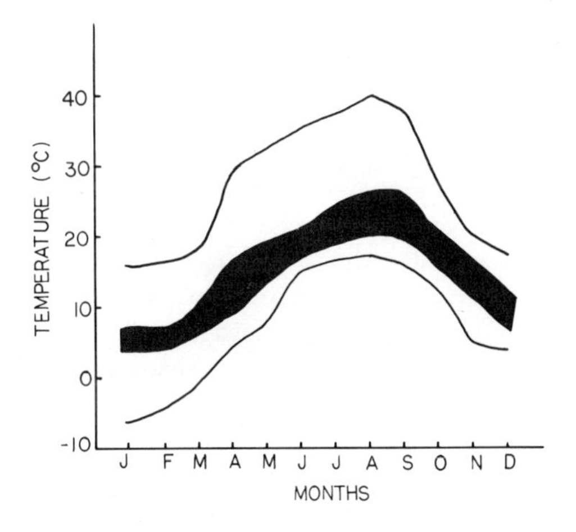

FIG. 3. A comparison of ambient and burrow temperatures in the high desert (1600 m) near Las Vegas, Nevada. The upper and lower lines represent monthly minimum-maximum averages of air temperature recorded 5 cm above the ground. The shaded area represents minimum-maximum averages of a Dipodomys microps *burrow.*

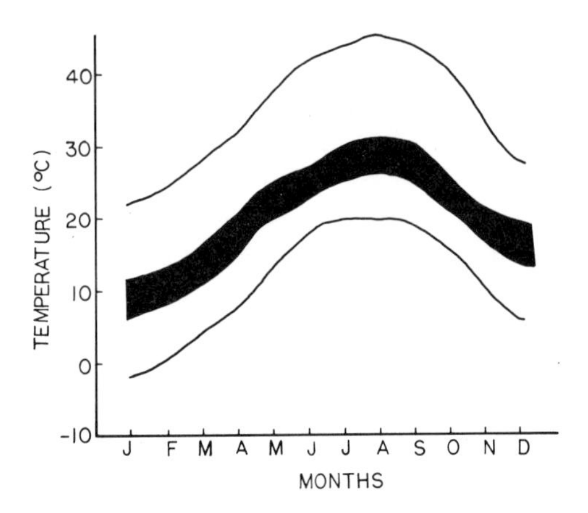

FIG. 4. A comparison of ambient and burrow temperatures in the low desert (650 m) near Las Vegas, Nevada. The upper and lower lines represent monthly minimum-maximum averages of air temperature recorded 5 cm above the ground. The shaded area represents minimum-maximum averages of a Dipodomys merriami *burrow.*

ACTIVITY PATTERNS

Most small mammals inhabiting deserts are nocturnal. Nocturnality is highly developed in all bats, and diurnal activity occurs only when bats are disturbed in their day roosts. With the exception of ground squirrels and chipmunks, all of the North American desert rodents are chiefly nocturnal (Fig. 5). Boreal and tropical species that inhabit marshes and grassy areas utilize runways. This group, although commonly considered nocturnal, exhibits varying degrees of diurnal behavior and is here termed partially diurnal. In addition, pocket gophers exhibit activity both night and day as they are almost completely fossorial and seldom above-ground.

Activity patterns in both diurnal and nocturnal mammals may be greatly modified by environmental features. The effect of ambient temperature on activity patterns of diurnal ground squirrels is shown in Fig. 6. Ground squirrels tend to be inactive at either low or high temperatures. Nocturnal activity patterns for two species of desert bats as related to ambient temperature are shown in Fig. 7. As illustrated by these figures avoidance of desert extremes by small mammals is achieved through fairly definite activity periods, which occur during the most favorable periods of the day or night. Seasonal inactivity or dormancy is discussed later.

MOBILITY AND MIGRATION

Bats are the only small mammals that are highly mobile, rivaling birds in their ability to fly long distances. This unique characteristic allows them to utilize

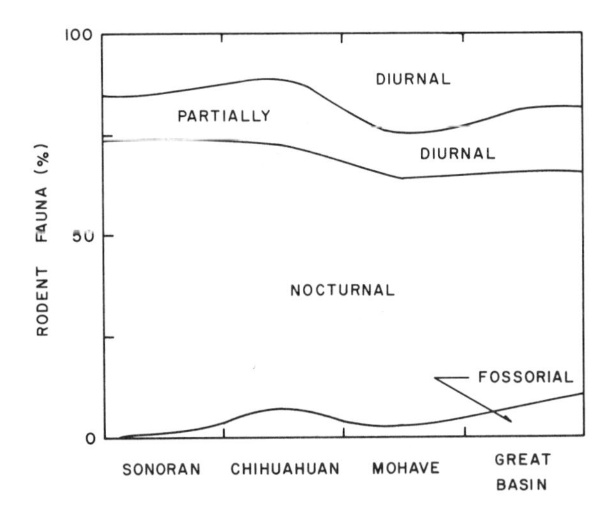

FIG. 5. The relative importance of different daily activity patterns among the rodent fauna of the North American deserts.

shelters, food, and water, which may be widely scattered and localized. For example, water and forage areas may be some miles distant from day roosts.

Most bats migrate to avoid stressful winter and summer conditions. Some of these migrations involve long distances. Populations of *Tadarida brasiliensis*, for example, which inhabit Texas, Oklahoma, New Mexico, and Arizona, spend the winter in Mexico, many wintering as far south as tropical Mexico. Tropical species that enter the southern deserts are largely found there only in summer months. Numerous north-temperate species move southward in the winter, whereas others move several to hundreds of miles to suitable winter hibernals (9). Some bat populations are year-round residents in certain areas of the southwest and apparently do not migrate. In southern Nevada two species are active throughout the year (10). Additional studies will probably reveal activity during winter months for additional species in warmer desert areas.

DIET

Since free water is scarce or lacking in the desert, food is an important source of water. Food resources are both sparse and variable due to a low and erratic rainfall.

Bats which inhabit north-temperate zones, including deserts, are predominantly insectivorous. As flying insects become inactive and absent as a food source, bats either migrate or hibernate (9).

Most desert rodents are opportunistic in their food habits. Seeds are relatively

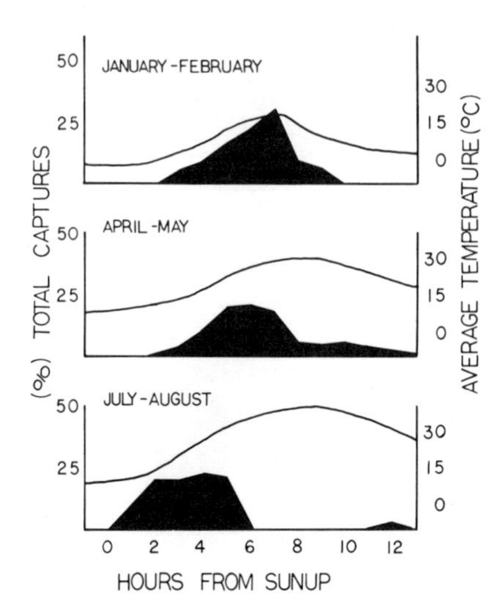

FIG. 6. The relationship between ambient temperature and activity pattern of a diurnal ground squirrel, Ammospermophilus leucurus.

available throughout the year, while most green vegetation and arthropods are seasonal. Hence most rodents utilize seeds as a stable food source and obtain other foods as they are seasonally available. Some rodents, e.g., kangaroo rats, are able to subsist on seeds throughout the year. However, there is evidence that succulence in the form of green vegetation is needed for successful reproduction (11, 12). Because of their higher water requirements ground squirrels and pack rats must rely on some succulence in the diet throughout the year (13, 14).

Physiological Adaptations

Though the physiological problems confronting small desert mammals are many and varied, most are related to temperature extremes, water scarcity, and food shortages. Physiological adaptations that accompany and complement the previously described modes of behavior are discussed in relation to these three basic environmental problems. The physiological responses of small mammals to desert environments include (a) adjustment of "physical" insulation by changing of fur conditions, i.e., smooth or fluffed; (b) adjustment of "physiological" insulation by altering body temperature and peripheral vascularity of the skin; and (c) adjustment of metabolic rate. These factors and their interactions determine the animal's body temperature T_b at a given ambient temperature T_a. The alteration in insulation and metabolic rate in response to changes in environmental conditions are discussed separately.

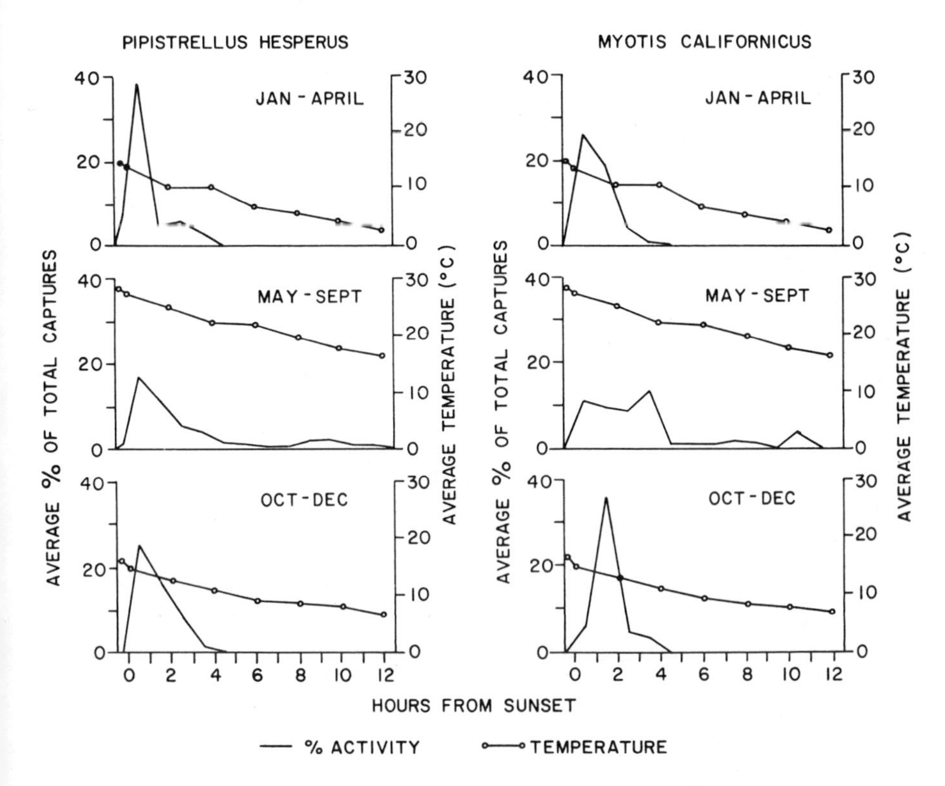

FIG. 7. The relationship between ambient temperature and activity patterns of two nocturnal bats (10). Activity (–); temperature (○).

INSULATION

In general, thermal conductance C equals the rate of heat loss at a given T_a divided by the temperature differential between the animal and its environment. The reciprocal of C is insulation, which means that high C represents a low insulation (15). For practical purposes, C can be estimated from the equation $M = C(T_b - T_a)$, where M is the metabolic rate. Therefore if M exceeds C, heat storage increases resulting in hyperthermia. Thus to prevent a lethal rise in T_b, M may be decreased and/or C increased; one or both may occur.

Differences in insulation among endotherms are considered an adaptation to the environment (15). This interpretation is reasonable, since insulation reduces both the rate of heat loss in cold environments and the rate of physical heat gain in hot environments. Available data on insulation of small desert rodents indicate a trend toward increased insulation. Comparisons of measured C to expected C calculated from the equation $C = 1.0\ W^{-0.5}$ derived by Morrison and

Ryser (16), show that most desert rodents have lower values than expected. In a study on *Peromyscus* from arid and mesic environments, it was concluded that species that live in the desert have two opposing conductive adaptations (17). One adaptation is low C that reduces the inward flow of heat from exposure to a radiant source. This is achieved by "physical" insulation, i.e., increased thickness of the fur. Heat loss under high T_a is facilitated by an increase in C probably attained by letting the fur mat or "break," supplemented by salivation as an emergency measure to facilitate further heat loss. It has been proposed that high C is advantageous in absence of radiation from sun and ground, whereas low C is beneficial in presence of radiation (17).

Both tropical and north-temperate zone bats have a higher C than that predicted from body weight (18). This high C may be related to the tropical origins of the order Chiroptera (19) accompanied by little natural selection for increased insulation in ancestoral bats evolved under stable tropical conditions.

METABOLIC RATE

It is generally accepted that M of fasting and resting mammals within the thermoneutral zone is inversely related to the weight of the animal to the 0.74 power (20). In their classic studies on adaptations of various birds and mammals to arctic and tropical environments, Scholander and associates introduced the concept that M of a homeotherm is phylogenetically nonadaptive to external conditions but is determined by body size irrespective of the climate (15). Recent studies indicate that the concept of nonadaptive M is not applicable to small desert rodents. Data shown in Fig. 8 indicate that M of most desert rodents is lower than expected as calculated from the equation derived by Brody (20).

Two mechanisms have been suggested to explain the lower M of desert rodents–behavioral response and higher body fat content (21, 22). These views cannot be accepted, since M represents minimal values in resting state within the thermoneutral zone. In addition, assuming that fat is metabolically inert, in order to produce a 40% reduction in M, an animal would have to increase its weight in fat by 40%. Thus a 35-gm rat would weight only 21 gm if it were fat free. Recent studies on Merriam's kangaroo rat have shown that total body fat content is very low (23, 24). Therefore it appears that a lower M is an adaptive mechanism that is probably molded by climate due to metabolic chemical "thermosuppression." In deserts a reduced M is advantageous in promoting a lower endogenous heat load and in conserving water and energy.

The M of tropical bats appears to be related to food habits (18). Species that feed on nectar, fruit, or animal matter have an M equal to or greater than that expected from Brody's equation, except in the largest species. In contrast, insectivorous bats, both tropical and north-temperate, have a M significantly lower than predicted. Therefore a reduced M appears to be a characteristic of bats that rely on an unpredictable food source.

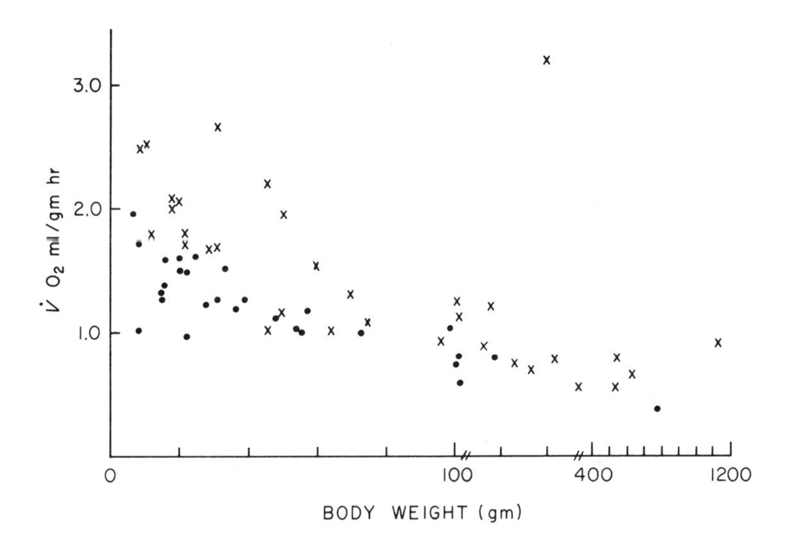

FIG. 8. A comparison of metabolic rate at thermoneutrality for desert (•) and nondesert (X) rodents. Comparative data are from P. Alttman, ed., "Biological Handbook: Respiration" (1971) and from our laboratory.

BODY TEMPERATURE

Within the zone of thermoneutrality most desert rodents have a T_b of 37°C or lower, whereas most nondesert rodents have a T_b of 37°C or above (Fig. 9). Scholander and associates suggested that if any adaptation occurred in T_b, arctic mammals should have lower T_b reducing $\Delta t(T_b - T_a)$ and tropical species have higher T_b increasing Δt permitting a greater dependence on physical heat-loss avenues (15). McNab and Morrison opposed this concept because the upper lethal temperature is only slightly above the normothermic level. Therefore when an animal is exposed to a high T_a any increase in T_b will have a lower base thus reducing the danger of hyperthermia (17). In their study the T_b of eight subspecies of *Peromyscus* was measured after a 2-hr exposure to a T_a of 38°C. The lowest T_b were for desert populations, and the highest were for mesic. This significantly lower T_b of desert subspecies represents an increased ability to maintain a small temperature differential during periods of high T_a.

The lower critical temperature (T_{lc}) has been estimated as the intercept between the slope of the line representing the increment of M in cold and the line describing M in the thermoneutral zone. Since Scholander and associates pointed out that the T_{lc} and insulation are related (15), some workers tend to use it as an index of thermoregulation (17). However, such index should be questioned since T_{lc} is related to C, M, and T_b. The upper critical temperature (T_{uc}) is the temperature where the physical thermoregulatory mechanisms can

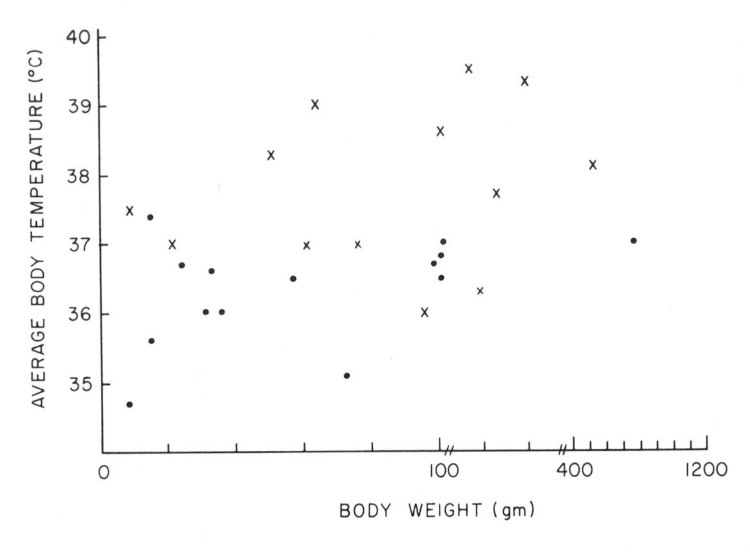

FIG. 9. A comparison of the average body temperature at or near thermoneutrality for desert (•) and nondesert (X) rodents. Comparative data from P. Alttman, ed., "Biological Handbook: Metabolism" (1968), and from our laboratory.

no longer maintain a normal T_b and M. The T_{uc} for most desert and nondesert rodents is similar; however, a study using four desert populations of *Peromyscus* showed that the T_{uc} was 1.4°C higher than in three mesic populations (17). These workers have suggested that a higher T_{uc} would be a beneficial adaptation, specifically for conservation of water. However, the behavioral burrowing that results in a moderate microclimate makes it unlikely that an increase of 1.4°C in the T_{uc} can be considered as a critical adaptive feature.

Temperate zone bats are considered to have a highly variable and low T_b when inactive (25). This generalization has been modified by McNab (18) who found that the mean T_b of nectarivorous, frugivorous, and carnivorous bats varied between 35° and 39°C, regardless of phylogenetic relationships. Additionally, insectivorous bats have a T_b range of 26°-35°C. Virtually all of the vespertilionid bats, which exhibit both daily torpor and hibernation, fall into this latter group.

HETEROTHERMY

Heterothermy refers to the phenomenon during which a homeotherm permits its T_b to drop to within a few degrees of T_a while retaining its ability to generate enough heat to restore its T_b to the normothermic level. Heterothermy occurs seasonally or daily involving varying periods of time. A reduced T_b is an adaptive mechanism in both winter and summer. In the summer a lower T_b *is associated* with reduced M resulting in less heat to be dissipated, thus conserving water. In the winter, a low T_b results in reduction of energy expenditure at a time when energy resources are limited. Time and space limitations preclude little more than a passing reference to physiological adjustments in seasonal hibernators and estivators (3, 26).

Accordingly, we shall limit our discussion to species that demonstrate short periods or daily hibernation. Many desert rodents, particularly the family Heteromyidae, exhibit periodic hibernation, which is commonly induced by shortage of food and low T_a (27, 28). There is recent evidence that this phenomenon occurs spontaneously in some *Perognathus* (29). We have similar evidence in our laboratory for *Dipodomys microps.* A reduced T_b and M is advantageous during periodic shortages of food as survival time on limited food supply is lengthened. The combination of food shortage and lowered T_a induces more extended periods of hibernation.

Temperate-zone bats inhabiting the American deserts normally exhibit daily hibernation and under seasonal cold either migrate or hibernate. A few rely on short periods of hibernation interrupted by foraging activity when nights are not too cold (10, 30).

WATER METABOLISM

A discussion of thermoregulation is not complete without simultaneous consideration of water metabolism. Water balance is a function of water intake and water loss, the latter is related to heat loss. In general, small desert mammals cannot afford to use water for thermoregulation because of their large surface area. Bats readily obtain free drinking water, whereas desert rodents must obtain water from other sources. In this respect they may be divided into two classes—first, those utilizing succulent plants or body fluids of their prey and second, those using dry food, therefore subsisting primarily on metabolic water. As for the first class, animals usually are behaviorally adapted to consume cactus and other plants with ease and, physiologically, they are able to feed on halophytic plants by excretion of concentrated urine and tolerance to toxic substances (2). The alternatives available for the second class may involve reduction in pulmonary water loss, urine volume, and fecal water loss. A number of studies have shown that pulmonary water loss is reduced by the high vapor pressure in burrows and by a countercurrent heat exchange in the respiratory passages. It should be pointed out here that cutaneous water loss is insignificant in small desert mammals because insensible water diffusion is reduced by the animal remaining in a high vapor pressure burrow.

Production of concentrated urine is an important factor in water balance. The concentrating power of the kidney is frequently expressed as the ratio between

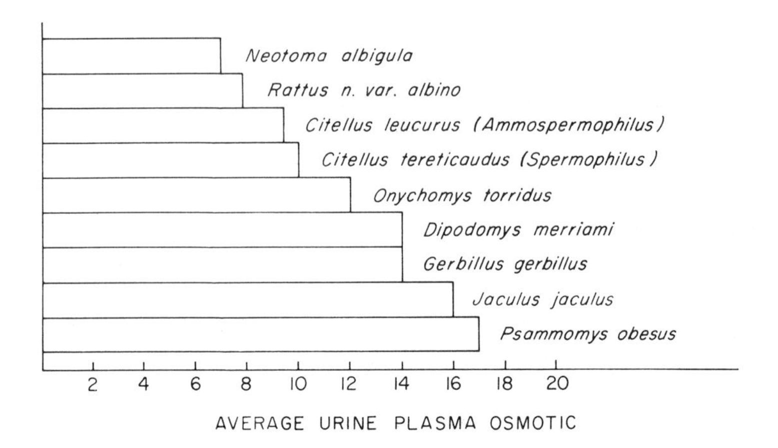

FIG. 10. A comparison of the U/P osmotic ratio of different desert rodents. Data on the white rat are shown for comparative purposes. Data taken from Schmidt-Nielsen (2) and J. W. Hudson, "Proceedings of the First International Symposium on Thirst in Regulation of Body Water." Pergamon, Oxford, 1964.

the concentration of urine and plasma, i.e., U/P ratio. Available data on the U/P ratio indicate that most desert rodents have kidneys with a high concentrating ability (Fig. 10). Reduction of fecal water loss is another avenue of water conservation. The few available data indicate that fecal water loss is reduced by reabsorption of water in the intestines and by reduction of the fecal bulk through coprophagy (2).

Another aspect of water balance is tolerance to dehydration. Desert rodents do not appear to possess any significant increase in the ability to tolerate dehydration as compared to nondesert forms. Comparative data are needed on the metabolism of water in dehydrated small rodents before any definite conclusion may be made.

Concluding Remarks

The various adaptive strategies found in small desert mammals involve complex interlocking patterns of behavior and physiology. Comparable behavioral and functional responses are found in small mammals inhabiting tropical and temperate environments. These responses, however, tend to be refined, sharpened and even exaggerated in small mammals that have evolved and adapted to desert conditions thus producing an array of evolutionary adaptations functioning synergistically in a particular species.

In general, behavioral adaptation is a primary feature of avoiding the rigors of desert environment. This is achieved by burrowing, using day roosts, nocturnality, mobility, and regulation of activity peaks to correspond with the most favorable T_a. Although these behavioral responses may be employed to the fullest, physiological adjustments are needed to complement them. Physiological adaptations include interactions of T_b, C, and M. Generally M is reduced and C may be reduced or increased depending on the species and its surrounding environment. In addition heterothermy is employed by many desert mammals to tailor their energy expenditure to properly fit food shortage, heat, cold, and aridity.

In conclusion, it is hoped that this review will stimulate further experimentation on physiological and behavioral characteristics of mountain, desert, and mesic populations of the same species. Data on water requirements and utilization rate in hydrated and dehydrated desert and nondesert mammals are needed to explain the poor adaptation to heat tolerance exhibited by most small desert mammals.

References

1. Schmidt-Nielsen, K., and Schmidt-Nielsen, B. (1952). Water metabolism of desert mammals. *Physiol. Rev.* **32**, 135.
2. Schmidt-Nielsen, K. (1964). "Desert Animals, Physiological Problems of Heat and Water." Oxford Univ. Press, London and New York.
3. Hudson, J. W., and Bartholomew, G. A. (1964). Terrestrial animals in dry heat: Estivators. *In* "Handbook of Physiology" (Amer. Physiol. Soc., D. B. Dill, ed.), Sect. 4, pp. 541-550. Williams & Wilkins, Baltimore, Maryland.
4. Chew, R. M. (1965). Water metabolism of mammals. *In* "Physiological Mammalogy" (W. V. Mayer and R. G. Van Gelder, eds.), Vol. 2, pp. 43-178. Academic Press, New York.
5. Bartholomew, G. A., and Dawson, W. R. (1968). Temperature regulation in desert mammals. *In* "Desert Biology" (G. W. Brown, ed.), Vol. 1, pp. 395-421. Academic Press, New York.
6. Axelrod, D. I. (1950). Evolution of desert vegetation in Western North America. *Carnegie Inst. Wash. Publ.* **590**, 215-306.
7. Hall, E. R., and Kelson, K. R. (1959). "The Mammals of North America," Vol. 1. Ronald Press, New York.
8. Bodenheimer, F. S. (1957). The ecology of mammals in arid zones. *In* "Human and Animal Ecology" p. 100. UNESCO, France.
9. Barbour, R. W., and Davis, W. H. (1969). "Bats of America." Univ. of Kentucky Press, Lexington.
10. O'Farrell, M. J., and Bradley, W. G. (1970). Activity patterns of bats over a desert spring. *J. Mammal.* **51**, 18.
11. Beatley, J. C. (1969). Dependence of desert rodents on winter annuals and precipitation. *Ecology* **50**, 721.
12. Bradley, W. G., and Mauer, R. A. (1971). Reproduction and food habits in Merriam's Kangaroo rat. *J. Mammal.* **52**, 497.

13. Bradley, W. G. (1968). Food habits of the antelope ground squirrel in southern Nevada. *J. Mammal.* **49**, 14.
14. Bradley, W. G., and Deacon, J. E. (1971). The ecology of small mammals at Saratoga Springs, Death Valley National Monument, California. *J. Ariz. Acad. Sci.* **6**, 36.
15. Scholander, P. F., Hock, R., Walters, R., and Irving, L. (1950). Adaptations to cold in arctic and tropical mammals and birds in relation to body temperature, insulation, and basal metabolic rate. *Biol. Bull.* **99**, 259.
16. Morrison, P. R., and Ryser, R. F. (1951). Temperature and metabolism in some Wisconsin mammals. *Fed. Proc., Fed. Amer. Soc. Exp. Biol.* **10**, Part 1, p. 93.
17. McNab, B. K., and Morrison, P. (1963). Body temperature and metabolism in subspecies of *Peromyscus* from arid and mesic environments. *Ecol. Monogr.* **33**, 63.
18. McNab, B. K. (1968). The economics of temperature regulation in neotropical bats. *Comp. Biochem. Physiol.* **31**, 227.
19. Griffin, D. R. (1958). "Listening in the Dark." Yale Univ. Press, New Haven, Connecticut.
20. Brody, S. (1964). "Bioenergetics and Growth." Hafner, New York.
21. Murie, M. (1961). Metabolic characteristics of mountain, desert and coastal populations of *Peromyscus. Ecology* **42**, 723.
22. Hayward, J. S. (1965). Metabolic rate and its temperature-adaptive significance in six geographic races of *Peromyscus. Can. J. Zool.* **43**, 132.
23. Yousef, M. K., and Dill, D. B. (1971). Responses of Merriam's Kangaroo rat to heat. *Physiol. Zool.* **44**, 33.
24. Yousef, M. K., and Dill, D. B. (1970). Physiological adjustments to low temperature in the kangaroo rat *(Dipodomys merriami). Physiol. Zool.* **43**, 132.
25. Stones, R. C., and Wiebers, J. E. (1965). A review of temperature regulation in bats (Chiroptera). *Amer. Midl. Natur.* **74**, 155.
26. Hoffman, R. A. (1964). Terrestrial animals in cold: Hibernators. *In* "Handbook of Physiology" (Amer. Physiol. Soc., D. B. Dill, ed.), Sect. 4, pp. 379-403. Williams & Wilkins, Baltimore, Maryland.
27. Tucker, V. A. (1966). Diurnal torpor and its relation to food consumption and weight changes in the California pocket mouse *Perognathus californicus. Ecology* **47**, 245.
28. Hudson, J. W. (1967). Variations in the patterns of torpidity of small Homeotherms. *In* "Mammalian Hibernation III" (K. C. Fisher *et al.*, eds.), pp. 30-46. Oliver & Boyd, Edinburgh.
29. Brower, J. E., and Cade, T. J. (1971). Bicircadian torpor in pocket mice. *BioScience* **21**, 181.
30. Bradley, W. G., and O'Farrell, M. J. (1969). Temperature relationships of the western pipistrelle *(Pipistrellus hesperus). In* "Physiological Systems in Semiarid Environments" (C. C. Hoff and M. L. Riedesel, eds.), pp. 85-96. Univ. of New Mexico Press, Albuquerque.

X

Principles of Adaptations to Altitude

U. C. LUFT

Although there are many astute descriptions of the unusual sensations and symptoms observed at high elevations in the mountainous regions of the world by early historians, particularly in connection with the conquest of South America by the Spaniards in the sixteenth century, it was probably Alexander von Humboldt, the famous explorer and natural scientist, who first recognized lack of oxygen as the essential cause of the loss of physical and mental powers at high elevations. After his unsuccessful attempt to reach the summit of Chimborazo (20,000 ft) in Ecuador he suggested in 1837 that on future expeditions to such altitudes one should carry a supply of "special air" enriched with oxygen. But it was not until Paul Bert's classic work "The Barometric Pressure" (1) that the effects of high altitude on the animal organism were approached with quantitative scientific methods in the laboratory and in the field. Now, for nearly 100 years the physiological alterations involved in adaptation to altitude have remained an object of sustained interest to physiologists, biochemists, and physicians. This has been stimulated by studies of permanent residents of high elevations of the earth, the ascent of the highest peaks on our planet by mountaineers during the last two decades, and also by the advent and spectacular progress in aviation culminating in successful manned space flights transcending the atmosphere of the earth. But perhaps the intrinsic reason for the intense and unabating scientific interest in the physiology and pathology of hypoxia is the fact that we are dealing with the single most critical and indispensable process in the warm-blooded organism, namely the uninterrupted supply of oxygen to the living cell and its utilization in the mitochondrion. All knowledge from the observation of the inherent ability of the healthy organism to survive and function at high altitudes contributes immeasurably to our understanding and proper management of many disease processes that jeopardize the vital oxygen transport system even in a normal environment at sea level. At this point it is appropriate to consider that the oxidative process can be diminished for a variety of reasons; it is meaningful to distinguish generally between *hypoxidation* (Table I) as the attenuation of aerobiosis associated with reduced energy requirement of the organism as in hypothermia, hibernation, the effect of hypokinetic drugs or hormonal disturbances such as hypothyroidism and *hypoxidosis* as the inability of aerobic processes to meet the undiminished or even increased energy requirements of the body due to deficiencies or malfunction of certain agents indispensable to the metabolic process. Even with an abundance of available oxygen cellular respiration may fail due to lack of energy substrates, blockage of enzymes by poisons such as cyanide, or the excessive accumulation of metabolites such as carbon dioxide. Undoubtedly, lack of oxygen is the most common cause of histohypoxia, which denotes a scarcity or lack of free oxygen molecules in the cellular environment.

This volume deals with adaptations to altitude where the primary concern is aerohypoxia, which is the general term we have proposed for reduced oxygen pressure in the inspired gas instead of the redundant expression hypoxic hypoxia frequently used in the past.

In considering the nature of hypoxia at altitude where the overall pressure head of oxygen between the air we breathe and the mitochondria of the cells is curtailed, there would appear to be essentially two modes for adaptive change available to the organism to alleviate this shortcoming. The first would be to minimize the impact of loss of oxygen pressure in the environment on its ultimate level in the cells. This involves the oxygen transport system of the body. A great majority of the most striking features of adaptation to altitude fall

TABLE I. Diminution of Oxidative Process

Condition	*Produced by*	*Characteristic*
Hypoxidosis Insufficiency of aerobic metabolism	Histohypoxia	Inadequate P_{O_2}
	Lack or dysfunction of enzymes Lack of energy sources Accumulation of metabolites	Inability to utilize O_2
Hypoxidation Reduced metabolic requirements	Hypothermia Hibernation Anesthesia Hypothyroidism	Adequate P_{O_2}

in this category and they contribute substantially toward maintaining a viable oxygen environment in the vital organs.

And yet even under optimal conditions for acclimation the integrated effects of all known adaptations of the oxygen transport system are not sufficient to preserve an oxygen environment in the cells anywhere near that at sea level. Therefore it has been postulated that a second mode of adaptation must exist whereby metabolic processes are modified so as to satisfy metabolic demands in spite of reduced availability of oxygen.

In order to assess the relative importance and also the limitations of the several factors improving O_2 transport in the acclimated individual we will follow the course of oxygen from ambient air to the tissue capillary and mixed venous blood in the form of a cascade over which the pressure head of O_2 is dissipated by a series of hindrances. In Fig. 1 the oxygen cascade at 22,000 ft is compared to that at sea level for the resting state. This altitude was chosen because mountaineers in the Himalayas have lived as high as this for weeks without untoward effects. The most striking feature in this comparison is that in spite of a loss of 92 mm Hg in the inspired P_{O_2} the mixed venous P_{O_2} is only 13 mm Hg less at 22,000 ft than at sea level. Apparently the flow of oxygen is greatly

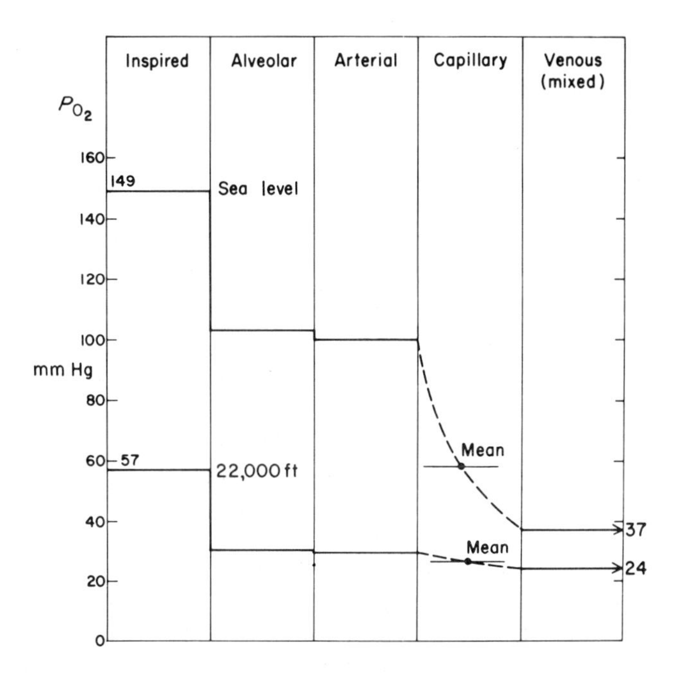

FIG. 1. The oxygen cascade from inspired air to mixed venous blood at sea level and for 22,000 ft.

facilitated at altitude. The inspired-alveolar P_{O_2} gradient, which varies directly with O_2 consumption and inversely with ventilation, is sharply reduced at altitude due to hypoxic hyperventilation. The second step in the cascade, the alveolar-arterial (A-a) gradient is normally attributable to a small amount of venous admixture via physiological shunts and unequal ventilation-perfusion relationships. This gradient becomes less at altitude because of the steeper slope of the O_2 Hb dissociation curve (2). During exercise, however, diffusion may become a limiting factor and increase the $A\text{-}aO_2$ gradient considerably (3). Diffusing capacity itself does not change at altitude in newcomers (3) but is reported to be higher in resident natives (4).

The most important step in the oxygen cascade from the capillary blood to the tissues is more difficult to assess because neither the mean capillary P_{O_2} nor the P_{O_2} in the tissues is directly accessible *in vivo*. However, a meaningful estimate of mean capillary O_2 pressure and the gradient to the cells where P_{O_2} may be less than 1 mm Hg can be obtained from knowledge of the arterial and venous P_{O_2}. The drastic cut in the arterio-mixed-venous O_2 gradient shown in our example for 22,000 ft is attributable for the most part to the inherent characteristics of the O_2- Hb dissociation curve first and second to an increase in O_2 capacity of the blood. Figure 2 shows the arterial and mixed venous points on two pairs of O_2- Hb and CO_2 dissociation curves, the lower one for O_2 with an O_2 capacity of 20 vol% and the higher one for 32 vol%, measured on climbers in the Himalayas after 6 weeks above 20,000 ft. The arteriovenous difference in O_2 tension for the same difference in O_2 content of 5 vol% would be 7 mm Hg instead of 58 mm Hg at sea level. These figures are calculated on the basis of Dill's standard O_2 dissociation curve for pH 7.4 (5), for a normal resting O_2 consumption of 0.250 liters/min, and a cardiac output of 5 liters/min. These premises are not strictly applicable to an individual at high altitudes. Original studies by Grollman (6) conducted on Pike's Peak and by Christensen and Forbes (7) on the expedition to Chile, and later by Asmussen and Consolazio (8), and others since, are in agreement that cardiac output tends to increase temporarily during the first week or so of acclimatization and returns to normal while other more permanent adaptations assert themselves. This is reasonable teleologically speaking since increased cardiac activity requires additional energy and reduces cardiac reserves for increased physical activity, as is well documented by the unusually high heart rates observed in unacclimatized individuals at altitude during exercise. Furthermore, the advantage to be gained by increased bloodflow is relatively small at altitude as compared to sea level, and it represents an expenditure with diminishing returns as shown in Fig. 3. A simple calculation of the increase in mixed venous P_{O_2} at rest obtained by a 50% increment in cardiac output gives 7 mm Hg at sea level but only 3.5 mm Hg at 15,000 ft. Raising cardiac output by further increments of 50% becomes less and less rewarding in terms of venous P_{O_2}, particularly at altitude. Incidentally, the increase in mixed venous P_{O_2} gained by a 50% increase in cardiac output is just about the same, namely 3-4 mm Hg, as is obtained by an increase in O_2 capacity of the blood from 20 to 30 vol% with a sea level cardiac output of 5 liters/min after adequate acclimation to 15,000 ft.

Another factor, which is not taken into account in Fig. 2, is a possible deviation of the position of the O_2- Hb dissociation curve which has been a matter of controversy ever since Barcroft *et al.* (9) suggested in 1922 that an increased affinity of hemoglobin for oxygen might be a major adjuvant to acclimation. Barcroft and his associates had found that fetal blood was greatly superior to adult blood of the same species in its affinity for oxygen, thus facilitating the intrauterine transfer of oxygen. He suggested that this might also

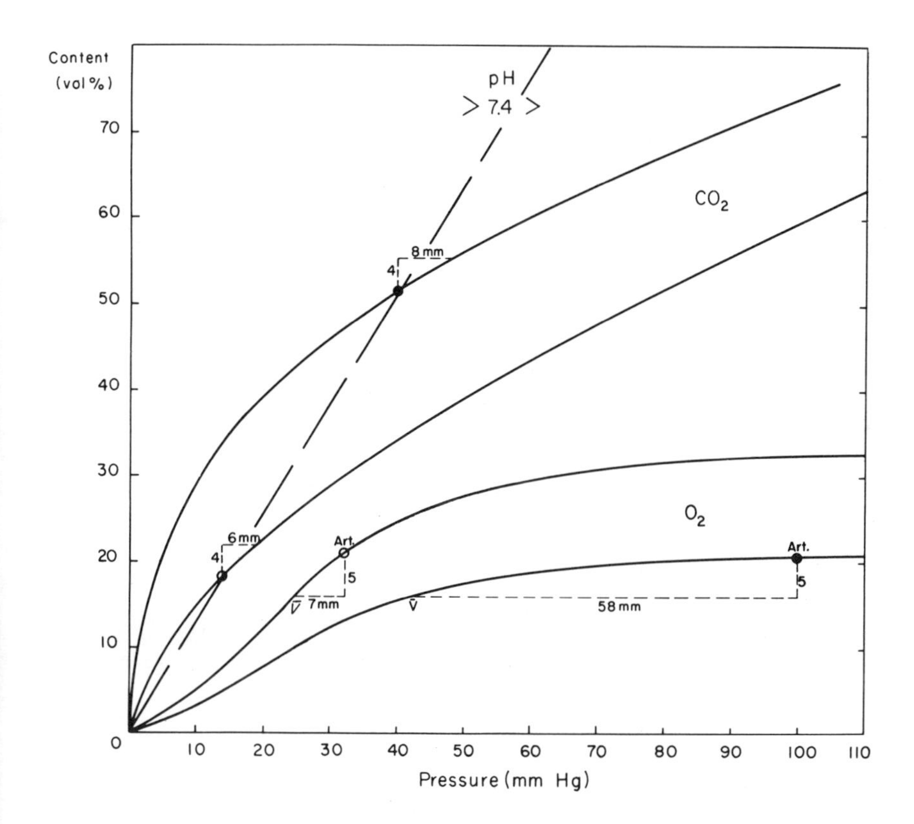

FIG. 2. Dissociation curves for oxygen and carbon dioxide with arterial and mixed venous points indicated for sea level (•) and for an individual acclimatized to 20,000 ft (○). The arteriovenous differences for O_2 (5 vol%) and for CO_2 (4 vol%) are assumed to be the same in both cases.

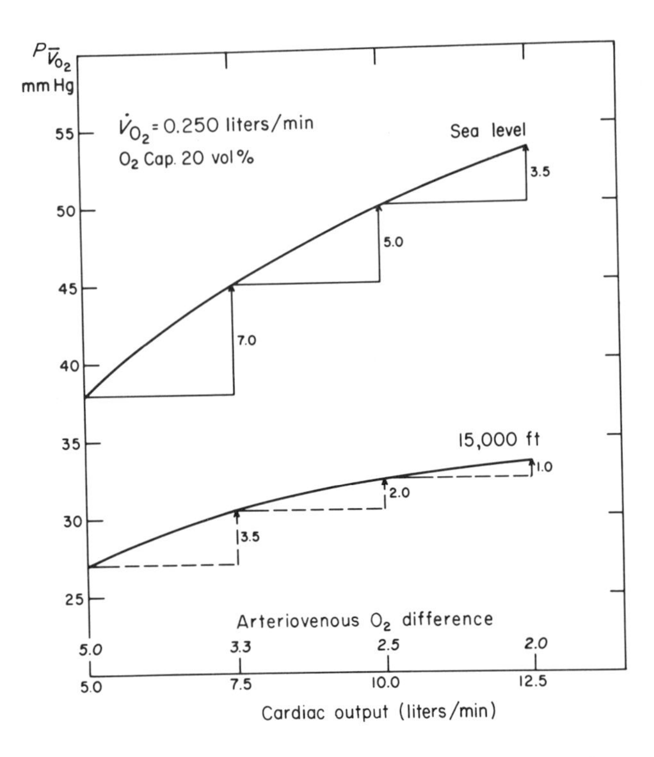

FIG. 3. Mixed venous oxygen tension as a function of cardiac output at sea level and at 15,000 ft. The oxygen consumption of 0.250 liters/min is the same for both curves. Oxygen capacity of the blood is 20 vol% at sea level and 30 vol% at altitude.

be true under conditions at high altitude where there is a lack of oxygen, in an address with the intriguing title "Everest *in Utero*." He went on to show from the results of his expedition to Peru (10) that there was a tendency for the dissociation curve to shift to the left, which could be explained partly by increased alkalinity during acclimation. Later studies by Dill and his associates in the mountains of Colorado (11) and in Chile (12) did not confirm this observation; on the contrary, their results showed if anything a decrease in O_2 affinity at altitude. Similar findings were made on natives at high altitude (13) in Morococha, Peru. Renewed interest in this fascinating subject has been generated in the past few years by the discovery (14, 15) that the concentration of certain organic phosphates, mainly 2,3 diphosphoglyverate (DPG) in the red blood cells and their binding with reduced hemoglobin can markedly reduce the O_2 affinity of hemoglobin and thus facilitate the unloading of oxygen in the tissues. Apparently the formation of DPG is stimulated whenever there is an increased amount of reduced hemoglobin in circulation, whereby DPG forms a complex binding with reduced hemoglobin thus diminishing the inhibitory effect it has on its own formation. Lenfant *et al.* (16) reported that a rapid and substantial rise in DPG occurs within 24 hr on ascent to high altitude (15,000 ft), and this was highly correlated with a shift to the right of the O_2- Hb dissociation curve measured as P_{O_2} at 50% saturation. The same phenomenon has been found in patients suffering from chronic hypoxemia as in congenital heart disease, chronic obstructive lung disease (17), and also noncyanotic heart disease, as well as during acute angina pectoris (18). Other studies performed on large numbers of residents in Leadville and Climax in Colorado (19) confirmed consistently higher levels of DPG in the blood than controls from the plains. There can be no doubt that not only the quantity of hemoglobin but also metabolic readjustments within the red blood cells can have a profound effect on safeguarding the oxygen supply to the tissues at high altitudes and in hypoxic states in general.

In an attempt to visualize the impact of the described changes in O_2 affinity of hemoglobin on overall O_2 transport we have drawn two pairs of O_2-dissociation curves for sea level (Fig. 4); one of each has a normal P_{50} of 26.5 mm Hg (A) and the other (B) is displaced to the right by 5 mm Hg as was found at 15,000 ft. Assuming a normal arteriovenous O_2 difference of 5 vol%, it is evident that at sea level with an O_2 capacity of 20 vol% (above) the mixed venous O_2 tension is increased by 6.0 mm Hg. The lower pair of curves represents a patient with anemia with a capacity of only 10 vol%. For the same arteriovenous difference, venous P_{O_2} is 13 mm Hg lower in this case but is improved by 4 mm Hg by the same shift to the right. Incidentally this example is very similar to actual observation on patients with pernicious anemia described by Dill in 1928 (20). On the graph in Fig. 5 we have plotted the same arteriovenous O_2 difference with normal O_2 capacity on the standard dissociation curve (A), and on curve B with the same shift to the right as on the previous graph. Arterial O_2 tensions are indicated corresponding to altitudes of 5000, 10,000, and 15,000 ft. The table at the foot of the graph summarizes the changes in mixed venous P_{O_2} and the effect of the decreased O_2 affinity as compared to sea level. The benefit of the shift becomes less with increasing altitude because the arterial O_2 content is progressively reduced on curve B so that at 15,000 ft venous O_2 tension is the same as on the standard dissociation curve. It would appear that this mechanism is of greatest advantage to the unloading of O_2 at the tissues when arterial oxygenation is high. If arterial saturation is below 70% the benefit is negligible. We have also considered the possibility that a shift of the curve to the right might be more effective in combination with a high O_2 capacity with polycythemia at altitude. Figure 6 gives a comparison of three pairs of dissociation curves with different O_2 capacities, one of each marked (A) with normal and the other (B) with reduced O_2 affinity. The arterial point on all curves is 40 mm Hg equivalent to 15,000 ft and the same arteriovenous O_2 difference has been applied. As shown in the table in Fig. 6, mixed venous O_2 tension on the upper curve A with an O_2 capacity of 30 vol% is more than 3 mm

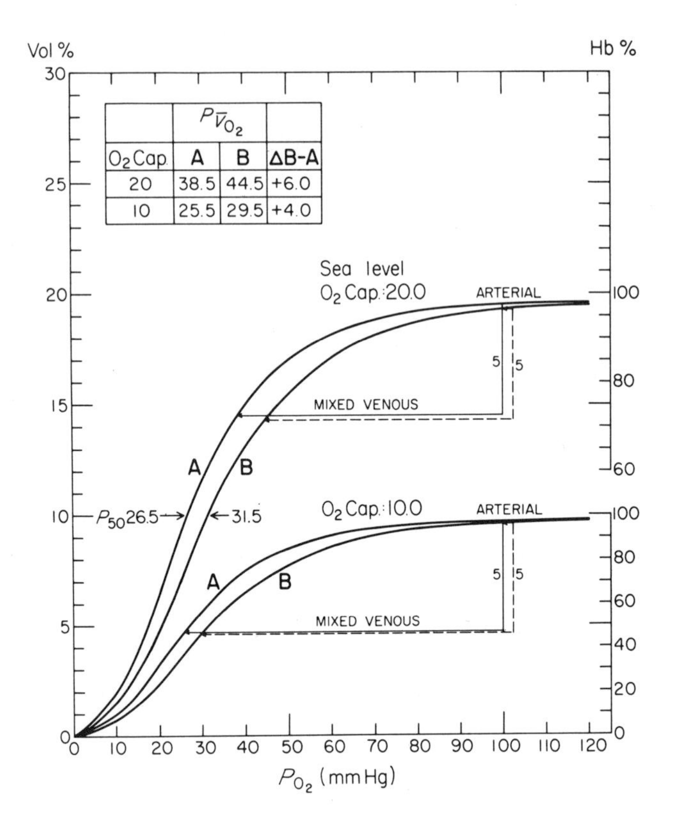

FIG. 4. The effect of reduced oxygen affinity of the blood on mixed venous P_{O_2} at sea level. Curves A have a P_{O_2} of 26.5 mm Hg at 50% saturation. (P_{50}), curves B of 31.5 mm Hg. Upper curves have a normal O_2 capacity of 20 vol%; the lower curves, 10 vol%.

Hg higher than on the curve A with a normal O_2 capacity of 20 vol%, while the shift to the right in curve B has no effect despite the increase in O_2 capacity. In the hypothetical case of anemia at 15,000 ft with an O_2 capacity of 10 vol% on the curves below, the mixed venous O_2 tension is actually slightly lower on curve B than A because of the change in slope at the foot of the dissociation curve.

Although the several adaptive mechanisms that we have considered so far contribute substantially toward safeguarding the integrity of aerobic metabolism at altitude, the tissues would still be functioning at an average oxygen tension considerably lower than at sea level as far as we can judge from the mixed venous P_{O_2}. We must assume that the critical P_{O_2} in the tissues, i.e., the oxygen tension below which the mitochondrial machinery can no longer operate properly, has been reduced. It has been established in studies on single

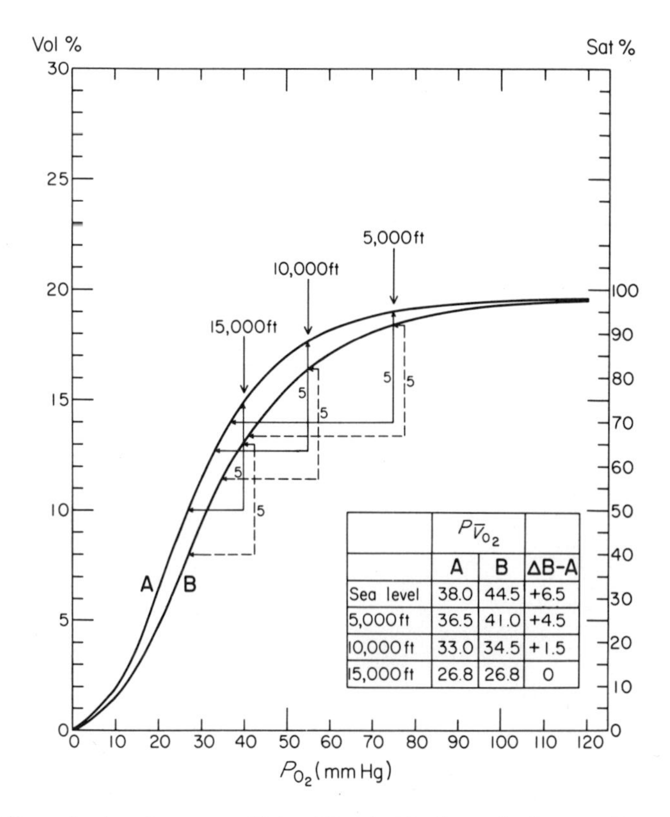

FIG. 5. Effect of reduced oxygen affinity (B) as in Fig. 4 on mixed venous P_{O_2} calculated for 5000, 10,000, and 15,000 ft with normal O_2 capacity showing less benefit to the unloading of oxygen at high altitudes.

mitochondria that their critical P_{O_2} is of the order of 1-3 mm Hg (21, 22) so that ordinarily there is a pressure head of at least 30 mm Hg for O_2 diffusion from the venous end of the capillary to the immediate vicinity of the mitochondria. The classic concept of the tissue cylinder surrounding a capillary originated by Krogh (23) serves to illustrate this point in Fig. 7 as modified by Thews (24) for the cerebral cortex under conditions at sea level. It is apparent that a drop in P_{O_2} of 10-15 mm Hg in the capillary blood at the venous end would seriously jeopardize the cells in the periphery of the cylinder at the venous end of the capillary, unless diffusion were facilitated. The most effective adaptation at this point would be a reduction of the diffusion distance by increasing the number of capillaries per unit of tissue. Convincing morphological evidence that this takes place in acclimatized animals is at hand from capillary counts in the cerebral cortex (25, 26), skeletal muscle (27), and myocardium

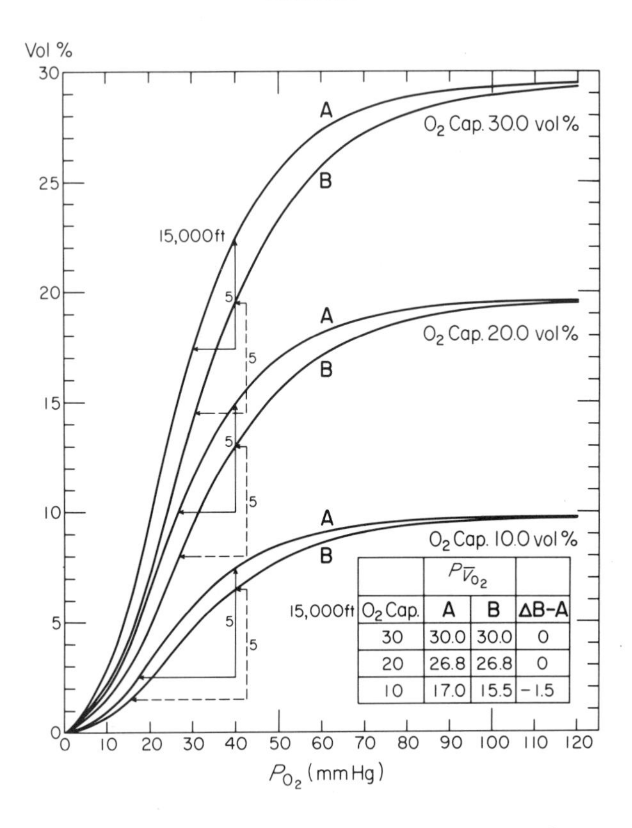

15,000ft	$P\bar{V}_{O_2}$		
O_2 Cap.	A	B	ΔB–A
30	30.0	30.0	0
20	26.8	26.8	0
10	17.0	15.5	−1.5

FIG. 6. Effect of high and low O_2 capacity of the blood on mixed venous P_{O_2} at an altitude of 15,000 ft with normal (A) and reduced (B) oxygen affinity as in Figs. 4 and 5.

(28). More recently a substantial increase in tissue diffusing capacity for carbon monoxide in animals exposed to the equivalent of 18,000 ft for 3 weeks has been interpreted as functional evidence of increased capillary density (29). Another important manifestation of adaptation at the tissue level is an increase in myoglobin content of heart and skeletal muscles first reported in animals at altitude by Hurtado *et al.* (30) and confirmed by numerous investigations since. Myoglobin acts not only as an acceptor for oxygen storage at low P_{O_2}, but it also enhances oxygen diffusion (31). A 40% increase in the number of mitochondria found in the hearts of cattle born and raised at 14,000 ft as compared to controls at lower elevations reported by Ou and Tenney (32) is further evidence for an active adaptive response in the tissues to hypoxia.

A great number of investigations have been devoted to the biochemical and enzymatic aspects of cellular adaptation to altitude. These have been reviewed in an admirable contribution by Barbashova (33).

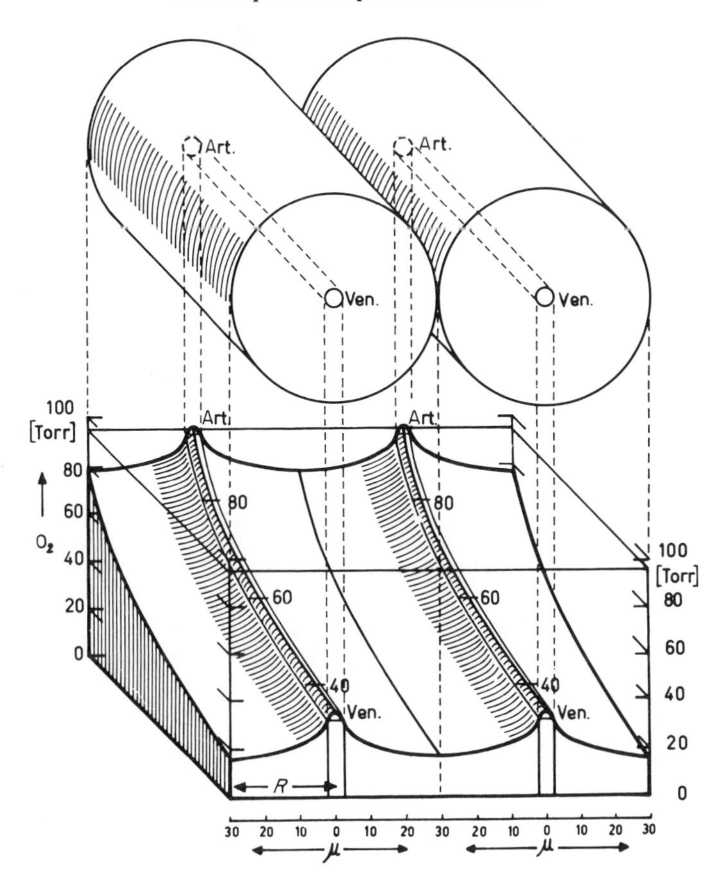

FIG. 7. Tissue cylinders surrounding two adjacent capillaries in the cerebral cortex indicating (below) oxygen tension profiles along the capillary and at the periphery of the cylinder. Note region most vulnerable to hypoxia in periphery at the venous end of the capillary. Calculated for sea level by Thews (24).

There is as yet no clear understanding whether there is a mechanism that reduces the energy requirements of the organism, analogous to the hypoxidation during hibernation, or whether the tissues maintain a normal level of energy production by acquiring the ability to utilize oxygen more efficiently under hypoxic conditions. On the other hand, there is strong evidence for increased participation of anaerobic metabolic activity (34). For instance, in their experiments with excised strips of heart and diaphragmatic muscle from animals acclimatized for 15 days to 22,000 ft, McGrath *et al.* (35) have demonstrated a significantly greater ability of the strips to function when they are stimulated under completely anaerobic conditions, compared to unacclimated controls. This phenomenon was enhanced in the presence of glucose and abolished by iodacetate, indicating that the increased tolerance to anoxia was due to a greater capacity for anaerobic glycolysis. These and other observations are difficult to reconcile with reports on reduced lactic acid production at altitude during and after exercise. On the International High Altitude Expedition to Chile in 1935 Edwards (36) found much lower lactic acid accumulation after exhausting exercise at altitudes ranging from 7200 to 17,400 ft than at sea level in members of the expedition. Similarly Reynafarje and Velasquez (37) found lactic acid concentrations only half as great after the same amount of work in natives in Morococha (14,900 ft) as in residents of Lima on the coast. Apparently this effect is present even at much lower elevations, since Cunningham and Magel (38) reported recently that lactate production in trained subjects after brief, exhausting exercise was 41% less after 4 days at altitude (7500 ft) and 51% less after 22 days than in controls in the lowlands.

All these observations point toward a lesser role of anaerobic energy production during exercise rather than increased glycolytic activity in the course of acclimatization. This may be a necessary adaptation, because the ability to cope with fixed acids in the blood is handicapped due to the loss of bicarbonate associated with the adjustments of acid-base balance to the hyperpnea at altitude (39) and the hydrogen ion concentration itself has been suggested as the factor limiting excess lactic acid production (40). This and many other facets of adaptation to altitude remain to be explored.

References

1. Bert, P. (1878). "La pression barométrique" (translation by M. A. Hitchcock and F. A. Hitchcock). College Book Co., Columbus, Ohio, 1943. Originally published by Masson, Paris.
2. Farhi, L. E., and Rahn, H. (1955). Theoretical analysis of the alveolar-arterial O_2 difference with special reference to the distribution effect. *J. Appl. Physiol.* 7, 699.
3. West, J. B., Lahiri, S., Gill, M. B., Milledge, J. S., Pugh, L. G. C. E., and Ward, M. D. (1962). Arterial oxygen saturation during exercise at high altitude. *J. Appl. Physiol.* **17**, 617.
4. Velasquez, T. (1956). "Maximal Diffusing Capacity of the Lungs at High Altitudes," Rep. No. 56-108. Sch. Aviat. Med., USAF, San Antonio, Texas.
5. Dill, D. B., and Forbes, W. H. (1941). Respiratory and metabolic effects of hypothermia. *Amer. J. Physiol.* **132**, 685.
6. Grollman, A. (1930). The effect of high altitude on cardiac output. *Amer. J. Physiol.* **93**, 19.
7. Christensen, E. H., and Forbes, W. H. (1937). Der Kreislauf in grossen Höhen. *Skand. Arch. Physiol.* **76**, 75.
8. Asmussen, E., and Consolazio, F. C. (1941). The circulation in rest and work on Mount Evans. *Amer. J. Physiol.* **132**, 555.
9. Barcroft, J., Binger, C. A., Bock, A. V., Doggart, J. H., Forbes, H. S., Harrop, G., Meakins, J. C., and Redfield, C. (1923). Observations upon the effect of high altitude

on the physiological processes of the human body carried out in the Peruvian Andes. *Phil. Trans. Roy. Soc. London, Ser. B* **211**, 351.

10. Barcroft, J. (1925). "Respiratory Function of the Blood. Part I. Lessons from High Altitudes." Cambridge Univ. Press, London and New York.
11. Dill, D. B., Edwards, H. T., Fölling, A., Oberg, S. A., Pappenheimer, A. M., and Talbott, J. H. (1931). Adaptations of the organism to changes in oxygen pressure. *J. Physiol. (London)* **71**, 47.
12. Keys, A., Hall, F. G., and Guzman Barron, B. S. (1936). The position of the oxygen dissociation curve of human blood at high altitude. *Amer. J. Physiol.* **115**, 292.
13. Aste-Salazar, H., and Hurtado, A. (1944). The affinity of hemoglobin for oxygen at sea level and at high altitudes. *Amer. J. Physiol.* **142**, 733.
14. Benesch, R., and Benesch, R. E. (1967). The effect of organic phosphates from the human erythrocyte on the allosteric properties of hemoglobin. *Biochem. Biophys. Res. Commun.* **26**, 162.
15. Chanutin, A., and Curnish, R. R. (1967). Effect of organic and inorganic phosphates on the oxygen equilibrium of human erythrocytes. *Arch. Biochem. Biophys.* **121**, 96.
16. Lenfant, C., Torrance, J., English, E., Finch, C. A., Reynafarje, C., Ramos, J., and Faura, J. (1968). Effect of altitude on oxygen binding by hemoglobin and on organic phosphate levels. *J. Clin. Invest.* **47**, 2652.
17. Lenfant, C., Ways, P. Aucutt, C., and Cruz, J. (1969). Effect of chronic hypoxic hypoxia on the O_2 Hb dissociation curve and respiratory gas transport in man. *Resp. Physiol.* **7**, 7.
18. Shappell, S. D., Murray, J. A., Nasser, M. G., Wills, R. E., Torrance, J. D., and Lenfant, C. J. M. (1970). Acute change in hemoglobin affinity for oxygen during angina pectoris. *N. Engl. J. Med.* **282**, 1219.
19. Eaton, J. W., Brewer, G. J., and Grover, R. F. (1969). Role of 2,3-diphosphoglycerate in the adaption of man to altitude. *J. Lab. Clin. Med.* **73**, 603.
20. Dill, D. B. (1928). Blood as a physiochemical system. VII. The composition and respiratory exchanges of human blood during recovery from pernicious anemia. *J. Biol. Chem.* **78**, 191.
21. Chance, B., Schoener, B., and Schindler, F. (1964). The intercellular oxidation-reduction state. *In* "Oxygen in the Animal Organism," p. 367. MacMillan, New York.
22. Lübbers, D.W., and Kessler, M. (1968). Oxygen supply and rate of tissue respiration. *In* "Oxygen Transport in Blood and Tissue" (D. W. Libben *et al.*, eds.), p. 90. Thieme, Stuttgart.
23. Krogh, A. (1936). "The Anatomy and Physiology of Capillaries." Yale Univ. Press, New Haven, Connecticut.
24. Thews, G. (1960). Die Sauerstoff-Diffusion im Gehirn. *Pfluegers Arch. Gesamte Physiol. Menschen Tiere* **271**, 197.
25. Mercker, H., and Schneider, M. (1949). Ueber Kapillveränderungen des Gehirns bei Höhenanpassung. *Pfluegers Arch. Gesamte Physiol. Menschen Tiere* **251**, 49.
26. Diemer, K., and Henn, R. (1965). Kapillarveränderungen in der Hirnrinde der Ratte unter chronischem Sauerstoffmangel. *Naturwissenschaften* **52**, 135.
27. Valdivia, E. (1958). Total capillary bed in striated muscle of guinea pigs native to Peruvian Mountains. *Amer. J. Physiol.* **194**, 585.
28. Cassin, S., Gilbert, R. D., and Johnson, E. M. (1966). "Capillary Development During Exposure to Chronic Hypoxia," Tech. Rep. 66-16. Brooks A. F. B., San Antonio, Texas.
29. Tenney, S., M., and Ou, L. C. (1970). Physiological evidence for increased tissue capillarity in rats acclimatized to high altitude. *Resp. Physiol.* **8**, 137.
30. Hurtado, A., Rotta, A., Merino, C., and Pons, J. (1937). Myohemoglobin at high altitude. *Am. J. Med. Sc.* **194**, 708.

31. Biörck, G. (1949). On myoglobin and its occurrence in man. *Acta Med. Scand., Suppl.* **226**, 133.
32. Ou, L. C., and Tenney, S. M. (1970). Properties of mitochondria from hearts of cattle acclimatized to high altitudes. *Resp. Physiol.* 8, 151.
33. Barbashova, Z. I. (1964). Cellular level of adpation. *In* Handbook of Physiology (Amer. Physiol. Soc.), Sect. 4, (D. B. Dill, ed.), Chapter 4, p. 37. Williams & Wilkins, Baltimore, Maryland.
34. Tappan, D. V., Reynafarje, B., Potter, V. R., and Hurtado, A. (1957). Alterations in enzymes and metabolites resulting from adaptation to low oxygen tensions. *Amer. J. Physiol.* **190**, 93.
35. McGrath, J. J., Bullard, R. W., and Komives, G. K. (1969). Functional adaptation in cardiac and skeletal muscle after exposure to simulated altitude. *Fed. Proc., Fed. Amer. Soc. Exp. Biol.* **28**, 1307.
36. Edwards, H. T. (1936). Lactic acid at rest and at work at high altitudes. *Amer. J. Physiol.* **116**, 367.
37. Reynafarje, B., and Velasquez, T. (1966). Metabolic and physiological aspects of exercise at high altitude. *Fed. Proc., Fed. Amer. Soc. Exp. Biol.* **25**, 137.
38. Cunningham, D. A., and Magel, J. R. (1970). The effect of moderate altitude on post-exercise blood lactate. *Int. Z. Angew. Physiol. Einschl. Arbeitsphysiol.* **29**, 94.
39. Luft, U. C. (1941). Die Höhenanpassung. *Ergeb. Physiol., Biol. Chem. Exp. Pharmakol.* **44**, 256.
40. Cerretelli, P. (1967). Lactacid oxygen debt in acute and chronic hypoxia. *In* "Exercise at Altitude" (R. Margaria, ed.), p. 58. Excerpta Med. Found., Amsterdam.

XI

Psychophysiological Implications of Life at Altitude and Including the Role of Oxygen in the Process of Aging

R. A. McFARLAND

Introduction

It is appropriate to include in this book a discussion of the influence of high terrestrial altitudes on behavior and performance. First of all, it is important to determine the physiological changes which occur in men and animals during acclimatization to high altitudes. It is equally as important, from the point of view of military operations, to analyze what effects these changes will have on human capacities for carrying out essential duties, as well as on the general well-being of men exposed to such environmental conditions.

Since the early development of aviation, both in flight at moderate and at extremely high altitudes, a great deal of research has been carried out on the ability of airmen to adjust to various conditions of oxygen want. Although the introduction of pressurized cabins has solved many problems, others have been created, such as the low humidity resulting from recirculating the cabin air, or a failure of the aircraft cabin giving rise to a sudden loss of pressure. Some of the experimental results from studies relating to hypoxia during flight can be applied to the problems of acclimatization of man to high terrestrial altitudes. In interpreting the findings of the effects of high altitude, however, it is always necessary to indicate the conditions under which the studies have been carried out.

Techniques of Measurement

Before summarizing some of the experimental findings relating to the effects of hypoxia on man, it is in order to consider the techniques of measurement used in such studies, both in regard to completed and to projected ones. Only brief reference will be made to this subject here, since a more thorough analysis of this area is presented in this book.

The subtle influence of hypoxia may often be masked by changes in the learning process or by "trying harder." Thus it becomes difficult to detect significant alterations in performance such as overall speed, number of errors, or total output. What appears to be required, therefore, are tests of performance which reflect deleterious effects of hypoxia in spite of attempts to overcome them, or tests which detect increased effort at the task. Recently, techniques have been developed in sensory and perceptual research that may be of value in developing measures of this kind.

Some of the primary features of such tests include (a) a high degree of sensitivity, so that small changes can be readily measured; (b) precision of the physical measurements involved in the test; (c) independence of the results from the degree of conscious or unconscious effort which may be exerted; (d) stability of the function during control experiments when the physiological stresses are not applied; and (e) for more complex performance, including insight and decision making, tests which involve the timing of perceptual responses, and information processing (1).

MEASURES OF SPEED AND ACCURACY COMBINED

One way of adjusting one's performance to offset the deleterious effects of altitude might be to sacrifice speed for accuracy or vice versa. Such trade-offs, however, need not indicate degradation in capacity. Until recently, there was no way of combining these two measures. With the development of information theory, a nonarbitrary way of combining measures of speed and accuracy into a single "rate-of-information-transmission" measure is available (2). In this way it is possible to determine the magnitude of a task-induced stress required to produce an environmentally interacting effect.

TESTS THAT DETECT AND SCALE INCREASED EFFORT

Two avenues of research offer tests of increased effort. One of these is concerned with capacity and/or peripheral attention. The measurement of performance on a primary task, with increased load on a secondary task, is a good example of this area of testing. The effects of environmental stress can be measured in terms of the steepening gradient of these spare capacity measures (3). The other area of research relates to physiological measures of arousal, of tension, and so forth. A number of techniques for measuring not only muscular tenseness, but also, central nervous system activity and even specific components of neural reaction to signals have been developed (4). Some of these measures appear to be promising for studying the deleterious effects of varying degrees of oxygen want.

In studies of the effects of noise or vibration on human performance there is evidence of strong interactions between such an environmental stress and task difficulty. The latter is very important in determining the amount of decrement that will appear under a given amount of stress (5). Task difficulty is a major variable. Failure to consider this fact has given rise to many of the discrepancies found in studies relating the influence of environmental stresses to performance. These studies have resulted in the hypothesis that the most demanding aspects of a complex activity are those that will be most sensitive to the effects of environmental stress. Furthermore the number of task elements affected by a given level of stress, and the extent of such impairment, depends on the total information processing load imposed by the task. Future studies of the effects of high altitude will undoubtedly include some of these techniques of measurement.

The Role of Oxygen in Biological Functions

Many years ago the physiologist Claude Bernard pointed out the role of oxygen in nervous functions. He stated that "the fixity of the internal environment is the condition of the free life," and that "all the vital mechanisms have only one object, that of preserving constant the conditions of life in the internal environment of the blood stream and fluid matrix of the body." He listed oxygen, water, temperature, and nutrient supplies (including salts, fats, and sugar) as important and necessary constants that must be controlled to free the organism from the limitations of the external environment (6).

The physiologist Barcroft (7) was one of the first to ask the important question, "Freedom for what?" In his view, it was chiefly freedom for the activity of the higher levels of the nervous system, especially the cerebral cortex. Thus in man's evolutionary development the body gained constancy over oxygen, temperature, acid-base balance, blood sugar, water, and certain other organic factors. Haldane devoted a lifetime of research in demonstrating the principle of automatic regulation in the field of respiration, especially the way in which the respiratory center responds to variations in carbon dioxide and oxygen (8).

Finally, the experiments of Cannon (9) clearly demonstrated the way in which behavior is influenced by variations in the internal environment, such as asphyxia, cold, or hypoglycemia—or, on the contrary, how external stimuli, giving rise to fear, anger, pain, or rage, may set in motion the homeostatic processes of regulation by way of the sympathicoadrenal division of the autonomic nervous system. Thus the organism in its evolutionary development ultimately reached a stage at which the higher faculties could develop. The constancy of the internal environment is most exact in man, and it is in man that the free life reaches its highest development. Thereby, intelligence, imagination, insight, and skill could be set free for higher services. This is also true of the person who is growing older and any failure in the supply or control of oxygen in the body tissues, especially the central nervous system, results in marked deterioration. This is what gerontologists would regard as senescence (1).

Review of Experimental Findings in Short- and Long-Term Exposures

In recent years several extensive bibliographies relating to high altitude have been prepared by NASA, the USAF, the United States Army Research Institute of Environmental Medicine, and the Guggenheim Center for Aerospace Health and Safety at Harvard University (10). In the preparation of this paper, selected references have been reviewed and the experimental findings relating to sensory and mental functions plotted on graphs. Unfortunately only a few studies are available from prolonged residence at high terrestrial altitudes. Great caution must be observed, therefore, in extrapolating some of the findings reported here to the problems of military operations at high terrestrial altitudes.

SENSORY FUNCTIONS

In this review certain tests relating to vision will be emphasized, although several other psychophysical functions will be reported. Visual acuity and light sensitivity possess many of the desired qualities for precise measurement. The changes manifested by the visual mechanism when its oxidative processes are disturbed are of considerable magnitude. Also, the measurements can be made very accurately. Experimental subjects are not aware of the changes in their own visual sensitivity, and for this reason they cannot mask the impairment by exerting greater effort.

Effect of Altitude on Light Sensitivity

The curve in Fig. 1 shows the extent of the decrease in ability to see under low levels of illumination in relation to altitude and oxygen concentration (11,12).

Light sensitivity is impaired at altitudes as low as 4000-5000 ft. At 16,000 ft ability to see may be reduced to 50% of sea level performance. Hence, twice as much light would be required for perception of a given stimulus under these conditions.

Effect of Altitude on Visual Acuity

The effects of altitude on performance tests of visual acuity are shown in Fig. 2 (13).

The different curves represent testing at different intensities of illumination. At the higher illuminations there is little or no impairment until altitudes beyond 18,000 ft are reached. Under low illumination however, a decrease in the ability to resolve a given target is apparent at about 8000 ft. At 17,000 ft visual acuity is reduced to 50% of the sea level value. Thus a target subtending almost twice the normal visual angle would be necessary for resolution under this condition.

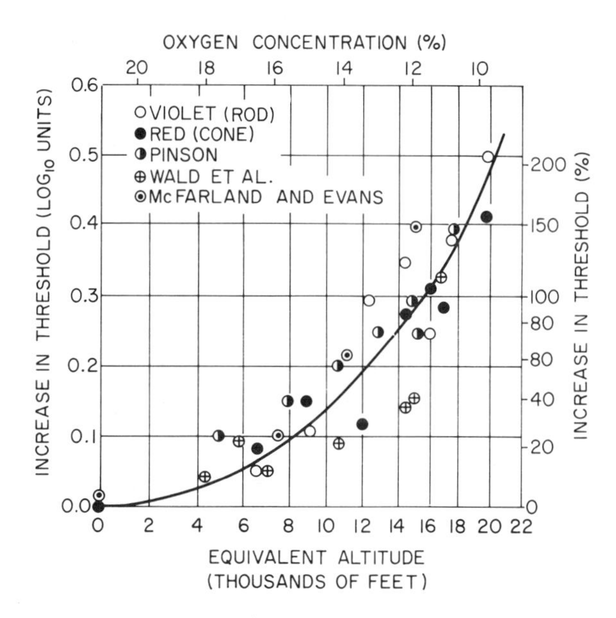

FIG. 1. The effect of altitude on light sensitivity (11, 12).

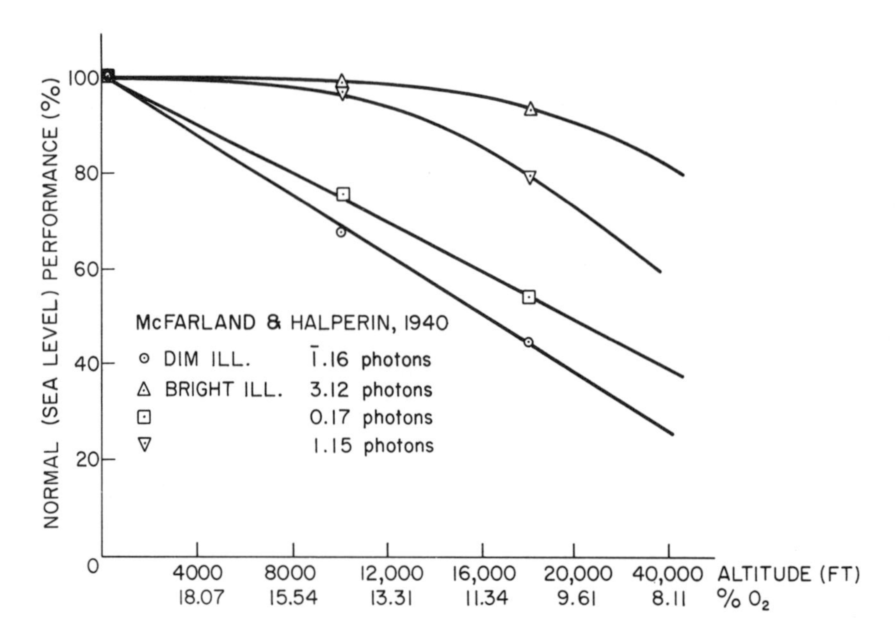

FIG. 2. The relationship between log visual acuity and log retinal illumination with normal air and at high altitude (13).

Effect of Altitude on Auditory Sensitivity

The relationship between auditory sensitivity and altitude has received less attention than the visual functions. Although the findings of the studies reported are not entirely in agreement, it appears that auditory sensitivity is highly resistant to hypoxia (14-16).

The curves in Fig. 3 (25a) show the average threshold in decibels at eight frequency levels for members of a high altitude expedition at sea level and at 17,500 ft. The threshold for normal hearing is also shown. Although the expedition group show a decrement from the normal hearing threshold at the upper frequencies, a further impairment is observed at an altitude of 17,500 ft.

Effect of Altitude on Memory

Data on the influence of altitude on memory functions have been summarized in Fig. 4 (17, 18). The curve clearly shows increasing impairment in memory with increasing altitude. The points plotted represent (a) a paired-word association task, (b) memory for pattern and position as measured by immediate recall, and (c) memory for pattern and position as measured by delayed recall. A

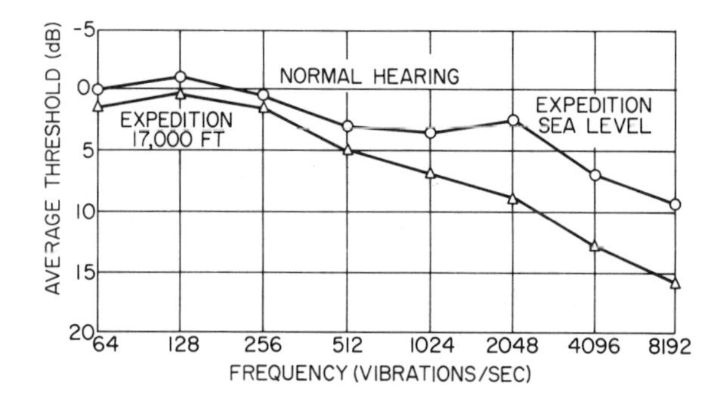

FIG. 3. The influence of high altitude on auditory thresholds at eight frequency levels (25a).

decrement in memory is apparent at 8000-10,000 ft and there is a more rapid decline after 12,000 ft.

The Relationship between Performance and Altitude

An attempt will now be made to interpret the above review of sensory and mental functions in relation to the altitudes at which losses in function occur–(a) the altitude at which the effects begin; (b) the altitude at which the relative change is a reduction in sensitivity of approximately 25%; and (c) the altitude at which performance is unacceptable (40%). It should be kept in mind that the graphs presented are based on limited data and, sometimes, inconsistent results.

The results of a series of visual functions are plotted in Fig. 5 (13, 18a-c). The curve for visual sensitivity clearly indicates the data for dark adaptation, or light sensitivity, provide the most sensitive test. The visual processes are initially impaired at 4500-5000 ft and reach unacceptable levels of impairment at 10,000-12,000 ft. This is an altitude well below that producing oxygen saturation levels that might be considered unacceptable along physiological lines.

The results of a series of mental tests, including pattern perception and decision making, are plotted in Fig. 6 (17, 33). It would appear that performance tests in these areas can be carried out at higher altitudes than for selected sensory tests. For example, such tasks can be performed successfully at altitudes in the neighborhood of 18,000-19,000 ft., where the blood oxygen saturation has dropped to 70-75%.

In Fig. 7 the relationship between arterial oxygen saturation and altitude is plotted for a series of performance tests in unacclimatized subjects (18d-h). From this chart one can observe the zones of minimal impairment for both

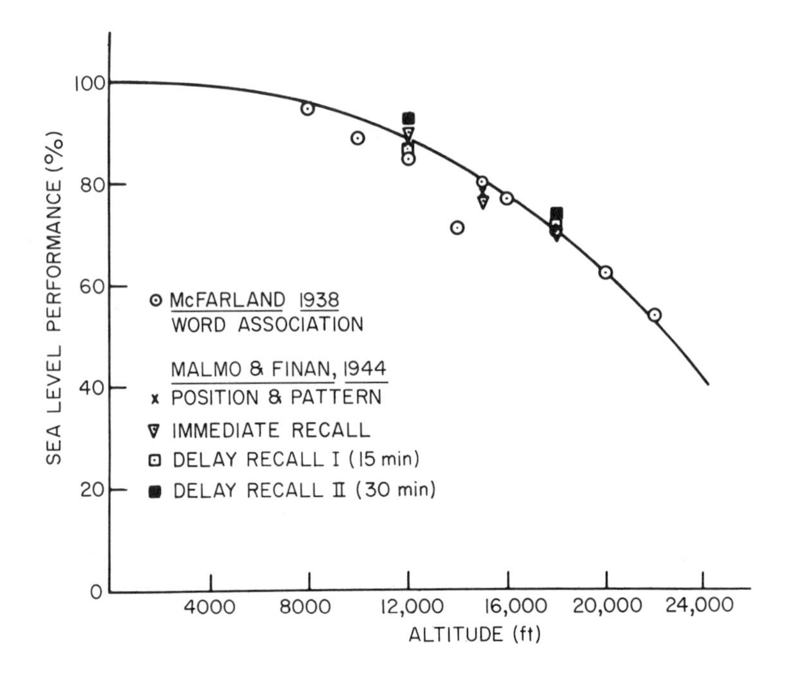

FIG. 4. The effect of altitude on selected tests of memory (17, 18).

sensory and mental functions. It is quite obvious from this figure that blood oxygen saturations between 70-75%, or altitudes of approximately 18,000 ft, represent the threshold for unacceptable performance, and higher altitudes might be expected to give rise to serious impairment.

Limitations of the Experimental Findings, with Special Reference to Gaps in Our Knowledge of Man's Ability to Acclimatize to High Terrestrial Altitude

In the brief outline below, reference will be made to some of the limitations and deficiencies in this field.

1. One of the most important problems deserving of consideration relates to the extent to which laboratory studies under decreased partial pressure, or total pressure, can be comparable to experiments carried out in mountainous areas during short or long time periods of acclimatization. It is obvious that additional studies are needed which would be considered comparable to military

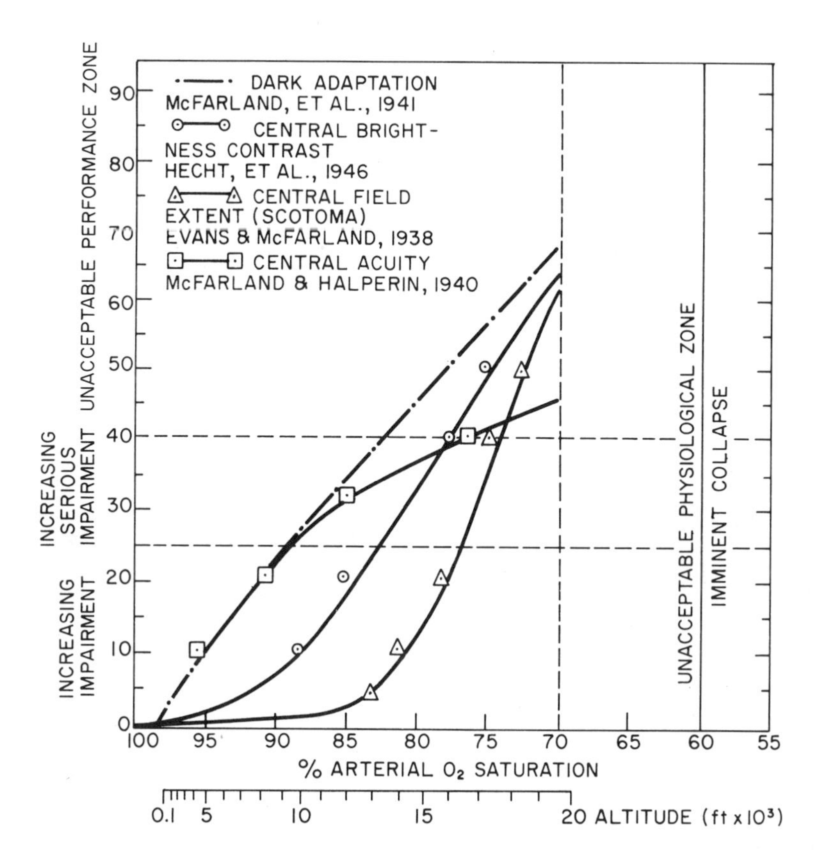

FIG. 5. The degree of impairment in performance of four visual functions in relation to oxygen level and altitude (13, 18a-c).

operations, if such are the major points of interest. The observation can be made with certainty that large numbers of people have become successfully acclimatized to very high altitudes. However, the rates of exposure and the specific altitudes that unacclimatized men in our military services can reach, for example, remain to be clearly defined.

2. The second major gap in our knowledge about the effects of high altitude concerns the limited variety of tests and the functions that have been measured. Information relating to judgment of distance, proprioception, vestibular function, and cutaneous responses is lacking. Little or no investigation has been made into these areas. The evidence for impairment in the extent of the peripheral field, a function that is of great importance to a soldier in the field, is meager and contradictory (19). No data were found on peripheral motion acuity, the ability to detect an object moving in the peripheral field.

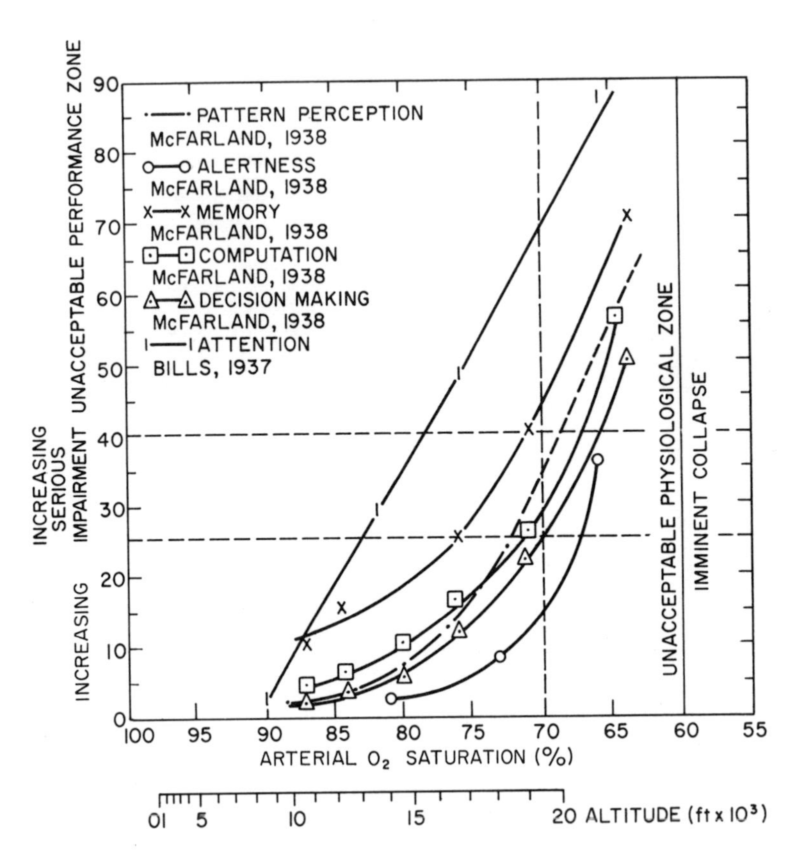

FIG. 6. The degree of impairment in performance of six mental tests in relation to oxygen level and altitude (17, 33).

3. One of the most difficult problems encountered in measuring performance concerns the control of both the learning process and motivation. It has been clearly established that it is necessary to have subjects well practiced if the effect of an environmental influence such as high altitude is to be successfully appraised. Also, motivation is an extremely difficult variable to control.

4. The variation in response from person to person has proved to be an important aspect of study of the effects of high altitude both in the laboratory at sea level and in mountainous areas. Individual variability is of great importance, therefore, and places great emphasis on initial selection of personnel, regarding age, physical fitness, and ability to perform, among many other factors.

5. The combined effects of altitude with other variables such as (a) drugs or sedatives for sleeping, (b) alcohol, (c) carbon monoxide from cigarette smoking, and (d) diet, are factors which must be considered.

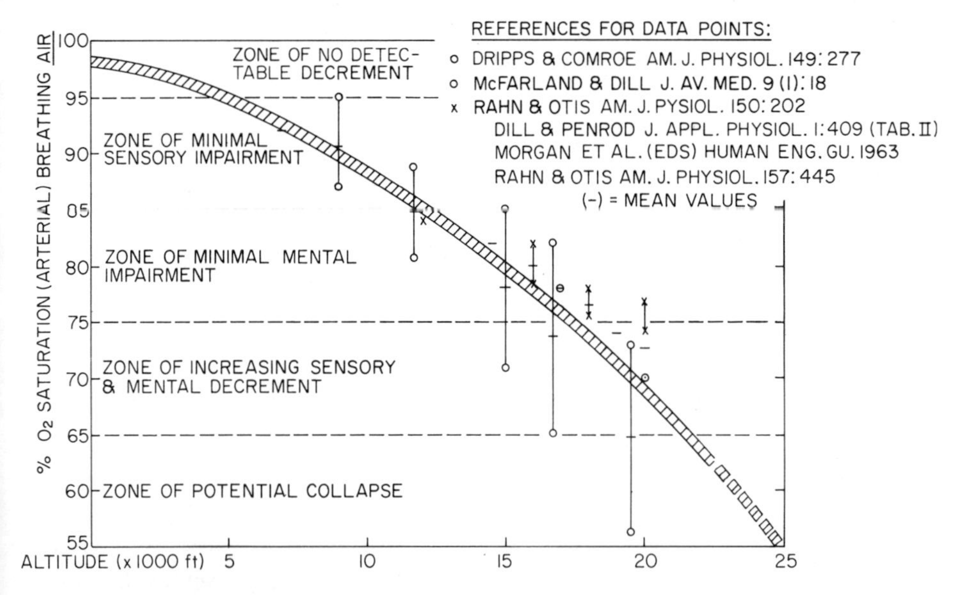

FIG. 7. The relationship between mean arterial oxygen saturation (%) and altitude for several performance tests in unacclimatized subjects (18d-h).

6. Other difficult factors to control are the influence of clothing, temperature extremes, and other variables that are apparent in the military situations. Most psychological tests and procedures, which are easily carried out under laboratory conditions, are difficult to adapt to field and military operations. Much remains to be done, therefore, in developing tests that would be relevant to military personnel operating at high terrestrial altitudes.

7. In reviewing the published literature in this field, it is quite obvious that the more recently developed methods of measurement for the study of environmental stresses, such as altitude, have not been widely used. Some of these more promising methods should be explored with respect to hypoxia as an important variable.

The Role of Oxygen in the Aging Process

The thesis that will be developed here as an important aspect of the aging process, as well as acclimatization to high altitude, concerns the diminished ability of the organism to transport or utilize oxygen in all tissues of the body. In searching for a comprehensive and basic approach to psychophysical relationships, one is impressed by the fundamental role of oxygen, primarily in the metabolism of nervous tissue. For example, the sensory and mental impairment that occurs in both normal and clinical subjects under experimental conditions of oxygen deprivation simulate very precisely the behavioral changes observed in the aging process. The major objective of this report will be to present experimental evidence of the important role oxygen plays in the functioning of the central nervous system (basic psychophysical relationships) and the implication, either direct or indirect, on the processes of aging.

CLINICAL EVIDENCE

The well-known degeneration of many higher functions, e.g., such as immediate recall, span of concentration, and insight, is accompanied by, if not caused by, an equally well-established impairment of cerebrovascular function. So are fainting, convulsions, and the senile or presenile degeneration of older persons. Similar observations are noted in patients whose arterial oxygen saturation falls below certain levels whether the illness be heart failure, pneumonia, or any toxic agent influencing the red blood cells (20).

An interesting clinical example may be seen in the rare disease progeria, a "galloping senescence" in which death occurs from old age within the first decade of life. Despite normal endocrine and cholesterol metabolism, advanced cardiac and cerebral arteriosclerosis develop, with corresponding impairment of oxygen transport (21). Even young patients with severe anemia show advanced symptoms of aging and senility. Chronic mountain sickness is a final example of an illness directly related to oxygen want. Usually referred to as Monge's disease (who first described it as a distinct clinical entity), the most prominent symptoms are loss of appetite, headache, weakness, disturbances of the special senses, distortion of mental functions, and transient spells of stupor and coma. Descent to sea level results in marked improvement or complete restoration of health (22).

PHYSIOLOGICAL EVIDENCE

The second line of evidence is in the association between physiological reactions common to old age and to hypoxia. Slowing of the basic resting frequency of the EEG is one such well-established observation (23). Impaired sensory functions such as dark adaptation, decreased auditory acuity for the higher frequencies, and increased reflex time also come to mind. According to Dill (24) "despite maintenance of good health, man experiences a gradual decline in function in vital organ systems, notably in those involved in the supply of oxygen to tissues."

The studies of mountaineers, aviators, and residents in mountainous areas present many illustrations of the similarity between healthy individuals temporarily suffering from oxygen want and those who are experiencing senility. This extends throughout the entire range of behavior, from the slowing of muscular responses to the impairment of sensation and the loss of memory and insight. Marked alterations in emotional control are common, and chronic fatigue states are symptomatic (25a-c). If mountaineers reach the summit of Mount Everest without an extra supply of oxygen, it will be a major physiological conquest as well as a mountaineering one, and such persons will undoubtedly be those in the younger age ranges.

A Comparison of Selected Psychophysical Tests Carried Out at High Altitude and in Relation to Age

INFLUENCE OF AGING ON LIGHT SENSITIVITY

Since it is known that changes in cerebral circulation and the oxygen transport system are involved in the aging process, studies were made of light sensitivity in relation to age. In the first experiment dark adaptation curves were obtained on approximately 200 pilots between the ages of 20 and 60 years with the Hecht-Shlaer apparatus (26). A correlation of –0.89 was found between age and final point of adaptation, i.e., the older subjects were poorer on the test (27). It was observed that the intensity of illumination at threshold levels must be approximately doubled for each increase of 13 years in age. This correlation appears to be one of the highest reported in the literature that relates a basic psychophysiological function to the aging process.

The findings in this study were so striking that it was decided to repeat the investigation on a more diversified group of subjects covering a wider age range (28). Using the same apparatus (29) 240 subjects, varying in age from 16 to 89 years, were given the test (see Fig. 8). The distributions were analyzed statistically, and mathematical models were derived for both rod and cone curves as a function of age. In Fig. 9 the results of this study are shown in relation to data obtained at high altitude.

One of the most striking features of the data was an indication of the degree of difference in sensitivity between the very old and the very young. For instance, at the second minute the youngest subjects were almost five times more sensitive than the most aged. And at the 40th minute the young were 240 times more sensitive than were the old. The degree of difference, then, between thresholds for the aged and for the young are not only relatively large, but are also geometrically progressive as a function of time. The results of the above analysis also demonstrated that rate of dark adaptation was found to be a function of age as well as of time (28, 30).

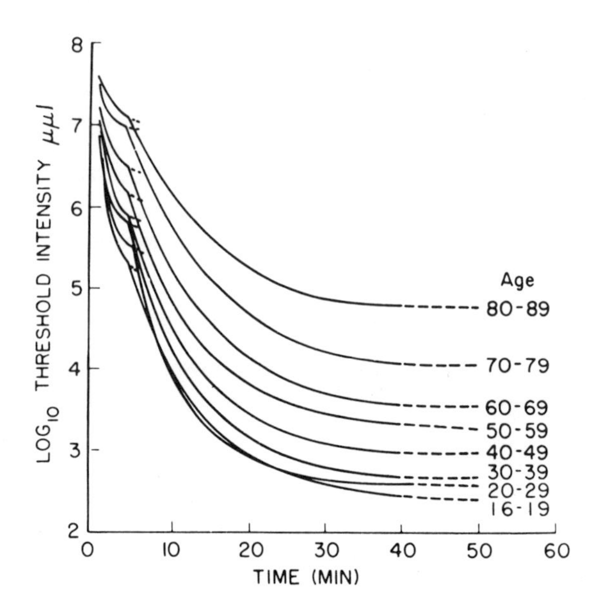

FIG. 8. Dark adaptation as a function of age for 240 subjects ranging in age from 16 to 89 years (28, 30).

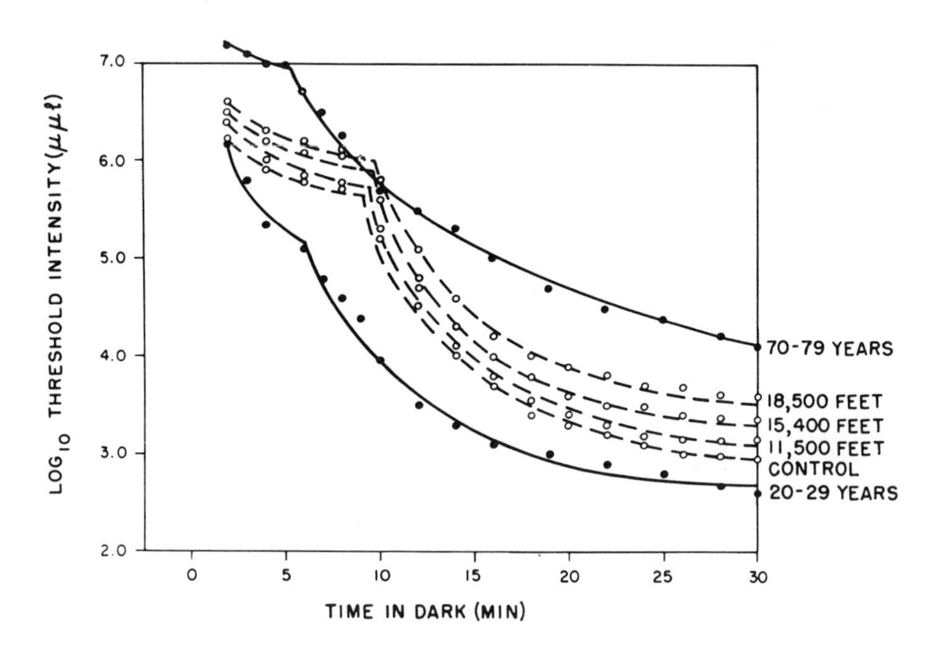

FIG. 9. Mean curves for dark adaptation obtained at simulated high altitude compared with those obtained for older and younger age groups (1, 28).

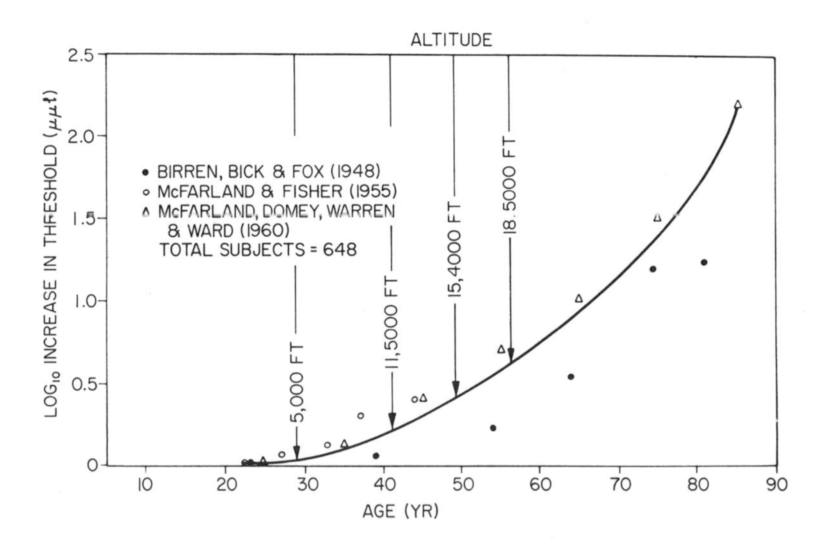

FIG. 10. The influence of age on dark adaptation, shown in relation to average values for the influence of altitude on dark adaptation (1, 27, 28, 30a).

In Fig. 10 the data from three different studies of dark adaptation in relation to age are shown in combination with the changes resulting from high altitude (1, 27, 28, 30a). The final points in the dark adaptation curves have been plotted for 648 subjects in relation to age. The data from the studies of high altitude were obtained using the same apparatus (Hecht-Shlaer Adaptometer) and similar experimental conditions.

INFLUENCE OF AGE AND HYPOXIA ON AUDITORY SENSITIVITY

The results from studies of another sensory field such as hearing are also of interest. The fact that auditory sensations are the last to disappear during loss of consciousness under various anesthetic gases is well known to surgeons. The balloonist Tissandier (31) was one of the first to report that, although vision was influenced early, he continued to hear sounds longer on the famous accidental ascent to 27,000 ft. Hearing was the first sense category to reappear when he regained consciousness upon returning to lower altitudes.

These observations have been confirmed in sudden and acute hypoxia experiments in chambers at sea level. Also, in the Chilean Expedition, audiograms were obtained on the scientists and the natives both at high altitude and at sea level. The mean audiograms for the ten expedition members are shown at sea level and at 17,500 ft in Fig. 11, related to data obtained on a large

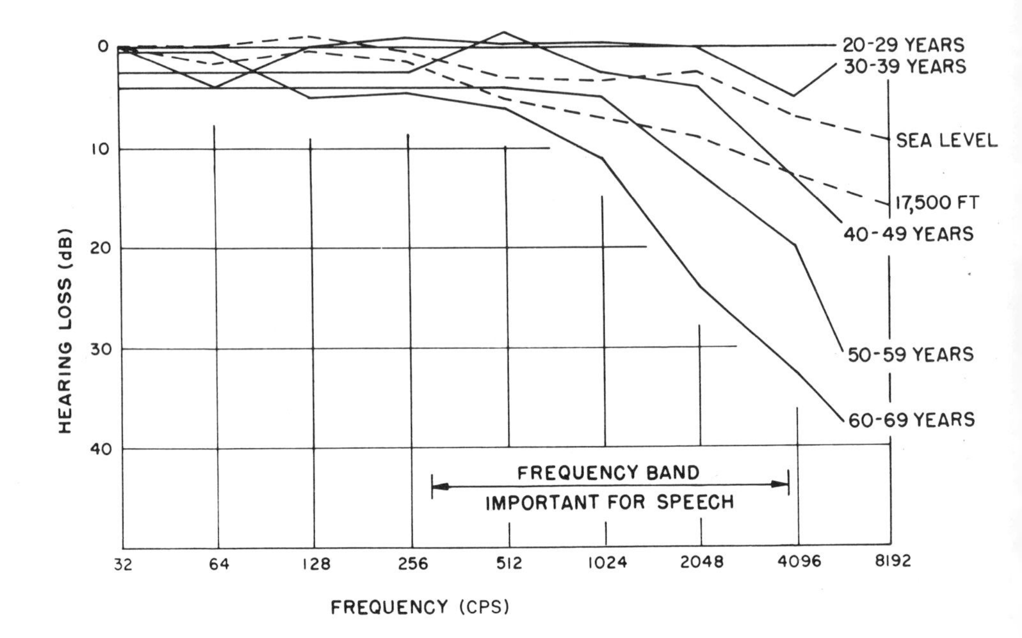

FIG. 11. The influence of high altitude on auditory acuity, shown in relation to the effect of age (1, 25, 25a-c).

number of subjects varying in age from 20 to 70 years. Of all the tests used this showed the least effects, but it is of interest that the higher frequencies were influenced by oxygen want to the greatest degree. In aging the same phenomenon is well known. The data from both studies are shown on the same graph. Why hearing is the least affected by hypoxia and why the higher frequencies are impaired more than the lower ones in both hypoxia and the aging process remains unexplained (1, 25a-c).

CHANGES IN COMPLEX MENTAL FUNCTIONS AT HIGH ALTITUDE AND WITH AGING

There are many striking similarities between the effects of high altitude and the changes that occur with aging in mental functions. The deterioration at altitude is characterized not merely as a general slowing up of mental functions and a greater amount of effort needed to carry out a task, but also with qualitative changes (32). Attention appears to fluctuate more easily and mental blocking is common (33). Calculations are unreliable, judgment faulty, and emotional responses unpredictable. The similarities between the effects of altitude and aging are very striking not only based on studies in altitude chambers, but also during airplane flights and expeditions to mountainous areas such as the Alps and the Andes (25, 25a-c). In the early history of aviation, airmen frequently forgot to check themselves in navigating and flew into mountains (34). Also mountaineers at high altitude have lost insight or have not been aware of their markedly deteriorated behavior (35).

The findings on tests of immediate memory under conditions of oxygen want and of aging are of considerable interest. It is well known that older persons are poorer in this function than younger ones. These reactions have been studied with the method of paired associates for meaningful words and nonsense syllables both at sea level and at high altitude (Fig. 12) (17, 18, 35a). The similarity of the results obtained are very striking indeed.

Other tests of mental functions in the aged, as well as at high altitude, have shown equally interesting results, although less objective methods of measurement are available. The loss of insight and judgment is quite similar in senility and in hypoxic states, especially if the oxygen saturation in the arterial blood drops below about 70-75%.

Thus far in this discussion of the problems of aging at high altitude the emphasis has been placed on the effects of oxygen want on the central nervous system. It has been shown that many of the most striking changes that occur in normal subjects at high altitude are precisely the ones that occur in persons as they grow older. Similarity in functional changes might not necessarily involve similar causal mechanism. In this case, however, the most important variable relates to oxidation, and it can be assumed that the underlying basis of the physiological and psychological changes in the aging process as well as at high altitude relates to an alteration in oxygenation of the tissues, whether it be a reduced supply, reduced circulatory delivery, or slower diffusion or utilization.

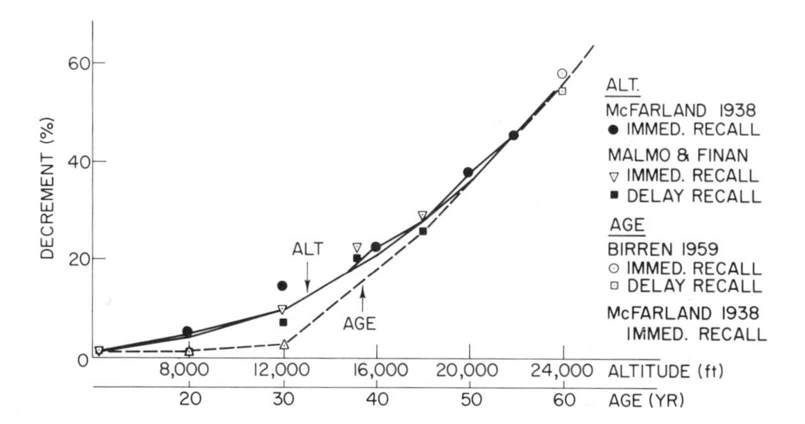

FIG. 12. The effects of altitude and aging on selected tests of memory (17, 18, 35a).

Consideration will now be given to the influences of oxygen want or high altitude on growth and development in the earlier as well as the later stages of the life cycle. In this way it might be possible to observe certain basic relationships between aging and altitude on the total organism.

Effects of High Altitude on Growth and Development

An attempt will now be made to observe some of the problems that have arisen at high altitude, not only during the period of gestation and at birth, but also during the period of development and growth through adolescence and maturity. The changes in barometric pressure and the amount of oxygen reaching the tissues, as well as its utilization, constitute one of the most important variables in responding to high altitude. Also, there is evidence to believe that there are basic changes in these processes during aging. Some of the evidence will now be presented relating to the effects of acute oxygen want and abnormalities in the newborn. This will be followed by a comparison of children born at high altitude with those at sea level in regard to body weight, and chest circumference. Comparison will also be made of selected variables, such as blood pressure, in adult males.

ACUTE OXYGEN WANT AND CONGENITAL ABNORMALITIES IN THE NEWBORN

Lack of oxygen is known to have a deleterious effect on the fetus. In experimental animals it can be shown that the amount of oxygen available at 15,000 ft to 20,000 ft altitude will produce a significant number of abnormalities in the offspring. The influences of oxygen lack during the first trimester or the organogenetic period is most critical. Ingalls *et al.* (36) exposed mice to altitudes of 27,000-30,000 ft during the complete reproductive cycle. They observed multiple skeletal deformities, cleft palate, mongolism, and other abnormalities that were attributed to the acute hypoxia in the mother and fetus. The basic physiology of the role of oxygen in fetal behavior was worked out by Barcroft (37).

Blindness in children (retrolental fibroplasia) is an example of a response of immature neural tissue to hypoxia. The effect may be induced by exposing a premature infant to an environment of high oxygen concentration and withdrawing him so rapidly that the physiological changes of acclimatization do not take place before irreversible changes occur in rapidly developing neural tissue such as the retina of the eye (38).

Studies of those who reside at high altitude present many interesting relationships in regard to the basic role of oxygen in all biological functions. When the Spaniards first migrated to the Andes there were so many miscarriages, stillbirths, and congenital anomalies that they had to take residence at lower altitudes. The Spaniards first settled in the city of Potosi, Bolivia (14,000 ft), about the year 1600, but the first child of Spanish parents was born there not until about 58 years later (22). During our expedition to Chile in 1935, it was observed that the highest permanent communities are at about 18,000 ft altitude. The natives can conceive their children at these altitudes, but the mothers go to lower altitudes to bear them (25, 25a-c).

Many of the current studies of developmental defects in children born and raised above 12,000-14,000 ft altitude tend to show a higher frequency of certain physical abnormalities. For example, in certain regions of the Andes the incidence of congenital malformations of the heart has been evaluated in two cities–one at sea level and the other at slightly over 12,000 ft. Twenty-five times more congenital abnormalities of the heart occurred in the city at 12,000 ft (39). See also studies in Lake County, Colorado, by Lichty *et al.* (40). Whether the life span of residents of high altitude is shorter has not been adequately investigated.

STUDIES OF BLOOD PRESSURE IN NATIVES AT HIGH ALTITUDE COMPARED WITH SUBJECTS AT SEA LEVEL OF COMPARABLE AGE AND RACE

During the International High Altitude Expedition to Peru and Chile in 1935, the author had an opportunity to obtain pulse rates and blood pressure under basal conditions. There were 35 miners in the group who lived at 17,500 ft and worked in an open pit, sulfur mining operation at approximately 19,000-20,000 ft altitude. The measurements were taken at the bedside of each miner before rising in the morning.

The mean values for pulse rates and for blood pressure for reclining, standing, and postexercise subjects are shown in Table I. The systolic and diastolic blood pressure for this group has been plotted in relation to age and that of United States males in Fig. 13 (25, 25a-c, 40a). The tendency for this group to have low values for all of the above variables is very striking (25, 25a-c). The critical ratios between the means of the miners and other groups studied are significant.

The extent to which these variables relating to pulse rate and blood pressure are due to selection, e.g., those miners who go to these great heights with higher values do not succeed in becoming acclimatized, is not known, and many other factors may be involved. Low values for pulse and blood pressure seem to be characteristic of the Andean residents however, and these observations offer interesting leads for further study in the field of cardiovascular physiology and disease.

TABLE I. Mean Pulse Rates and Blood Pressure Records for Miners and Expedition Members Taken at 17,500 ft Altitude and for Native Workmen at Sea Level (25, 25a-c)

	High altitude miners (17,500 ft)	*Native workers (sea level)*	*Expedition members (17,500 ft)*
Pulse rate			
Reclining	51 ± 5.5	64 ± 5.6	66.3
Standing	60 ± 6.2	74 ± 8.7	83.0
After exercise	75 ± 10.1	90 ± 10.1	100.4
Blood pressure			
Reclining	107/79	121/79	115/80
Standing	110/80	118/79	111/82

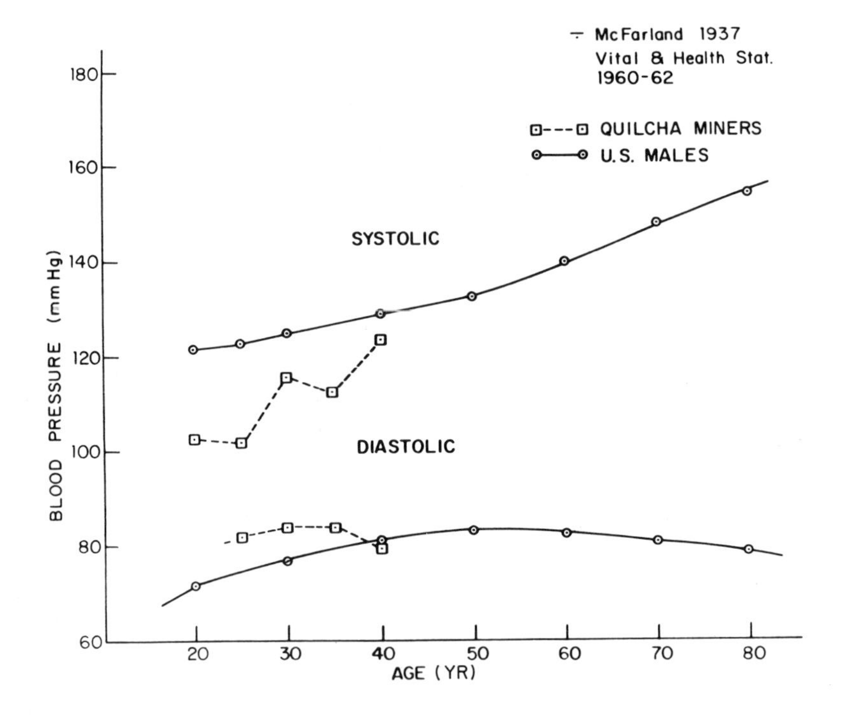

FIG. 13. The systolic and diastolic blood pressure for U.S. male population compared to miners living at 17,500 ft and working as high as 20,000-21,000 ft altitude in northern Chile (25, 25a-c, 40a).

ADAPTATION TO HIGH ALTITUDE IN RELATION TO AGE

During 1962 it was possible to carry out studies of six of us who were members of the original party of ten on the International High Altitude Expedition to Chile and Peru in 1935, or 27 years later. The respiratory minute volume in rest increases about as rapidly and to the same extent as in young men. Nevertheless, several of the six were slower to acclimatize than before as indicated by dyspnea on exertion, headache, Cheyne-Stokes breathing, and associated loss of sleep. In exercise, easy work on the ergometer was performed with the same respiratory minute volume as in 1935. As the grade of work was increased the minute volume increased more than in 1935; the oxygen consumption in peak performance was much less than in 1935. In general, it appeared that adaptation was slower than when we were 27 years younger but that this was not due to the respiratory system (41).

In regard to certain other variables, such as the rate at which members of the party showed an increase in hemoglobin, there appeared to be a distinct slowing. In general, despite the maintenance of good health we all experienced a gradual decline or slowness of function in the vital organ systems of the body, notably in those involved in the supply of oxygen to the tissues (24). See Table II for biochemical test results on natives and members of expedition.

Summary and Conclusions

Many observations on both normal and diseased subjects have indicated that there is a decrease in cerebral metabolic rates or oxygen consumption with age. In some instances, there is evidence of loss of cerebral tissue. Reduction in the total number of cells, for any reason, is therefore another cause for decreased oxygen uptake. According to the Himwiches, hypoxia may be a result rather than a cause of aging (42). However, in some diseased conditions hypoxia precedes the cell destruction. Whereas one would not wish to imply that oxygen is the sole factor in aging, as is the case in responding to high altitude, this variable is one of the most closely related and basic single factors. Multiple causality is undoubtedly involved in such a complex process as aging.

Several lines of evidence have been summarized that suggest a basic similarity among the kinds of behavior, both physiological and psychological, encountered in oxygen want at high altitude and in aging. In fact, some of the evidence indicates a cause and effect relationship, with hypoxia implicated as an etiological agent in senescence. A precautionary note should be added, however,

TABLE II. Biochemical Determinations for Eight of the Permanent residents (Quilcha Miners) 17,500 ft (401 mm Hg) Compared with the Means for the Ten Members of the Expedition at the Same Altitude (25c)

Subject	*Oxygen saturation (%)*	*Oxygen capactiy (vol %)*	*Red blood cells (million)*	*Arterial pH's*	*Arterial p_{CO_2} (mm Hg)*	*Arterial p_{O_2} (mm Hg)*	*Alkaline reserve*[a]
Quilcha miners							
Ala	83.0	29.8	–	7.45	26.0	47.8	38.8
Alc	78.3	29.2	6.85	7.35	31.1	47.3	34.6
Bas	70.2	30.7	7.09	7.36	31.7	34.4	36.0
Cam	74.8	32.8	7.48	7.34	30.9	42.3	33.0
Frz	84.6	24.6	6.15	7.38	22.8	52.0	34.0
Her	67.6	34.2	9.10	7.32	34.7	38.9	33.0
Mar	75.2	–	–	7.42	27.9	39.9	40.1
Tro	–	32.5	7.54	–	–	–	33.4
Mean	76.2	30.5	7.37	7.37	29.3	43.2	35.4
S.D.	6.3	3.3	0.89	0.08	3.7	6.7	3.0
Expedition members							
Mean	76.2	25.1	5.96	7.43	27.7	43.1	–
S.D.	4.9	1.9	0.66	0.03	3.4	3.4	–

[a]Total CO_2 of oxygenated blood at p_{CO_2} = 40mm. (cf. McFarland 25c, p. 214)

that living in an enriched oxygen atmosphere would not necessarily prevent aging or prolong life!

Although an analysis of the basic physiochemical mechanisms underlying these processes are beyond the scope of this presentation, in both cases, i.e., hypoxia and aging, there is a diminished availability, or utilization, of oxygen in the central nervous system. This may be due to either or both of the following causes of interference with normal metabolic processes within the individual cells: (a) a reduced rate of oxygen transfer to the cells or (b) inadequacy of cellular enzyme systems. Transference in turn may be inadequate because of reduced oxygen supply, reduced circulatory delivery, or slower diffusion.

It can also be concluded, in regard to the problems reviewed here concerning acclimatization to high altitude, that (a) this field of study offers a very basic approach to many aspects of the physiological mechanisms of the body; (b) a better understanding of cerebral functions can be obtained in analyzing the effects of hypoxia in both acclimatization to high altitude and in aging; and (c) many new approaches can be worked out to the more traditional areas in the fields of mental and emotional illness, growth and development in the earlier stages of life, and certain cardiovascular diseases.

References

1. McFarland, R. A. (1963). Experimental evidence of the relationship between ageing and oxygen want: In search of a theory of ageing. *Ergonomics* **6**, 339.
2. Broadbent, D. E. (1958). "Perception and Communication." Pergamon, Oxford.
3. Posner, M. L. (1966). Components of skilled performance. *Science* **152**, 1712.
4. Wilkinson, R. T. (1964). Changes in performance due to environmental factors. *In* "Symposium on Medical Aspects of Stress in the Military Climate," pp. 127-135. Walter Reed Army Med. Center, Washington, D.C.
5. Weisz, A. Z., Goddard, C., and Allen, R. W. (1965). "Human Performance under Random and Sinusoidal Vibration," AMRL-TR-65-209. Wright-Patterson Air Force Base, Ohio.
6. Bernard, C. (1878). "Leçons sur les phénomènes de la vie communs aux animaux et aux végétaux," Vol. I. Baillière et Fils, Paris.
7. Barcroft, J. (1938). "The Brain and Its Environment." Oxford Univ. Press, London and New York.
8. Haldane, J. S., and Priestley, J. G. (1935). "Respiration." Yale Univ. Press, New Haven, Connecticut. Second Edition.
9. Cannon, W. B. (1935). Stresses and strains of homeostasis. *Amer. J. Med. Sci.* **189**, 1.
10. Tune, G. S. (1964). Psychological effects of hypoxia: Review of certain literature from the period 1950 to 1963. *Percept. Mot. Skills* **19**, 551.
11. McFarland, R. A. (1953). "Human Factors in Air Transportation, Occupational Health and Safety." McGraw-Hill, New York.
12. McFarland, R. A., and Evans, J. N. (1939). Alterations in dark adaptation under reduced oxygen tensions. *Amer. J. Physiol.* **127**, 37.
13. McFarland, R. A., and Halperin, M. H. (1940). The relation between foveal visual

acuity and illumination under reduced oxygen tension. *J. Gen. Physiol.* **23**, 613.

14. Curry, E. T., and Boys, F. (1956). Effects of oxygen on hearing sensitivity at simulated altitudes. *Eye, Ear, Nose Throat Mon.* **35**, 239.
15. Klein, S. J., Mendelson, E. S., and Gallagher, T. J. (1961). The effects of reduced oxygen intake on auditory threshold shifts in a quiet environment. *J. Comp. Physiol. Psychol.* **54**, 401.
16. Klein, S. J. (1961). Effects of reduced oxygen intake on bone conducted hearing thresholds in a noisy environment. *Percept. Mot. Skills* **13**, 43.
17. McFarland, R. A. (1938). "The Effects of Oxygen Deprivation (High Altitude) on the Human Organism," Rep. 13. Dept. of Commerce, Bureau of Air Commerce, Safety and Planning Division, Washington, D.C.
18. Malmo, R. B., and Finan, J. L. (1944). A comparative study of eight tests in the decompression chamber. *Amer. J. Psychol.* **57**, 389.

18a. McFarland, R. A., Evans, J. N., and Halperin, M. H. (1941). Ophthalmic aspects of acute oxygen deficiency. *Arch. Ophthalmol.* **26**, 886.

18b. Hecht, S., Hendley, C. D., Frank, S. R., and Haig, C. (1946). Anoxia and brightness discrimination. *J. Gen. Physiol.* **29**, 335.

18c. Evans, J. N., and McFarland, R. A. (1938). The effects of oxygen deprivation on the central visual field. *Amer. J. Ophthalmol.* **21**, 968.

18d. Dripps, R. D., and Comroe, J. H., Jr. (1947). The effect of the inhalation of high and low oxygen concentrations on respiration, pulse rate, ballistocardiogram and arterial oxygen saturation (oximeter) of normal individuals. *Amer. J. Physiol.* **149**, 277.

18e. McFarland, R. A., and Dill, D. B. (1938). A comparative study of the effects of reduced oxygen pressure on man during acclimatization. *J. Aviat. Med.* **9**, 18.

18f. Rahn, H., and Otis, A. B. (1947). Alveolar air during simulated flights to high altitudes. *Amer. J. Physiol.* **150**, 202.

18g. Dill, D. B., and Penrod, K. E. (1948). Man's ceiling as determined in the altitude chamber. *J. Appl. Physiol.* **1**, 409.

18h. Morgan, C. T., Cook, J. S., III, Chapanis, A., and Lund, D. W., eds. (1963). "Human Engineering Guide to Equipment Design." McGraw-Hill, New York.

19. Halstead, W. C. (1945). Chronic intermittent anoxia and impairment of peripheral vision. *Science* **101**, 615.
20. McFarland, R. A. (1941). The internal environment and behavior. Part I. Introduction and the role of oxygen. *Amer. J. Psychiat.* **97**, 858.
21. Grossman, H. J., Pruzansky, S., and Rosenthal, I. M. (1955). Progeroid syndrome. Report of a case of pseudo-senilism. *Pediatrics* **15**, 413.
22. Monge, C. (1937). High altitude disease. *AMA Arch. Intern. Med.* **59**, 32.
23. Obrist, W. D. (1954). The electroencephalogram of aged adults. *Electroencephalogr. Clin. Neurophysiol.* **6**, 235.
24. Dill, D. B. (1961). "The Physiology of Aging in Man." The George Cyril Graves Lecture presented 6 February 1961, in the Department of Anatomy and Physiology, Indiana University, Bloomington, Indiana (unpublished).
25. McFarland, R. A. (1937). Psycho-physiological studies at high altitude in the Andes. Part I. The effects of rapid ascents by aeroplane and train. *J. Comp. Psychol.* **23**, 191.

25a. McFarland, R. A. (1937). Part II. Sensory and motor responses during acclimatization. *J. Comp. Psychol.* **23**, 227.

25b. McFarland, R. A. (1937). Part III. Mental and psycho-somatic responses during gradual adaptation. *J. Comp. Psychol.* **24**, 147.

25c. McFarland, R. A. (1937). Part IV. Sensory and circulatory responses of the Andean residents at 17,500 feet. *J. Comp. Psychol.* **24**, 189.

26. McFarland, R. A., Graybiel, A., Liljencrantz, E., and Tuttle, A. D. (1939). An analysis of the physiological and psychological characteristics of two hundred civil air line pilots. *J. Aviat. Med.* **10**, 160.
27. McFarland, R. A., and Fisher, M. B. (1955). Alterations in dark adaptation as a function of age. *J. Gerontol.* **10**, 424.
28. McFarland, R. A., Domey, R. G., Warren, A. B., and Ward, D. C. (1960). Dark adaptation as a function of age. I. A statistical analysis. *J. Gerontol.* **15**, 149.
29. Hecht, S., and Shlaer, S. (1938). An adaptometer for measuring human dark adaptation. *J. Opt. Soc. Amer.* **28**, 269.
30. Domey, R. G., McFarland, R. A., and Chadwick, E. (1960). Dark adaptation as a function of age and time. II. A derivation. *J. Gerontol.* **15**, 267.
30a. Birren, J. E., Bick, M. W., and Fox, C. (1948). Age changes in the light threshold of the dark adapted eye. *J. Gerontol.* **3**, 267.
31. Tissandier, G. (1875). Le voyage à grande hauteur du ballon "Le Zénith." *Nature (Paris)* **3**, 337.
32. McFarland, R. A. (1932). The psychological effects of oxygen deprivation (anoxemia) on human behavior. *Arch. Psychol., N.Y.* No. 145, p. 1.
33. Bills, A. G. (1937). Blocking in mental fatigue and anoxemia compared. *J. Exp. Psychol.* **20**, 437.
34. Barach, A. L., McFarland, R. A., and Seitz, C. P. (1937). The effects of oxygen deprivation on complex mental functions. *J. Aviat. Med.* **8**, 197.
35. Greene, R. (1957). Mental performance in chronic anoxia. *Brit. Med. J.* **1**, 1028.
35a. Birren, J. E., ed. (1959). "Handbook of Aging and the Individual, Psychological and Biological Aspects" Univ. of Chicago Press, Chicago.
36. Ingalls, T. H., Curley, F. J., and Prindle, R. A. (1950). Anoxia as a cause of fetal death and congenital defect in the mouse. *Amer. J. Dis. Child.* **80**, 35.
37. Barcroft, J. (1936). Fetal circulation and respiration. *Physiol. Rev.* **16**, 103.
38. Szewczyk, T. S. (1952). Retrolental fibroplasia. Etiology and prophylaxis. *Amer. J. Ophthalmol.* **35**, 301.
39. McCreary, F. J. (1959). "The Child and His Environment." A series of four lectures presented at the Public Health Institute, Department of Health, Health Branch, British Columbia.
40. Lichty, J. A., Ting, R. Y., Bruns, P., and Dyar, E. (1957). Studies of babies born at high altitude. I. Relation of altitude to birth weight. *Amer. J. Dis. Child.* **93**, 666.
40a. National Center for Health Statistics (1960-1962). Blood pressure of adults by age and sex. Vital and health statistics. *U.S. Pub. Health Serv., Publ.* **1000**, Ser. 11, No. 4, 1-40.
41. Dill, D. B., Forbes, W. H., Newton, J. L., and Terman, J. W. (1964). Respiratory adaptations to high altitude as related to age. *In* "Relations of Development and Aging" (J. E. Birren, ed.), Chapter 5, pp. 62-73. Thomas, Springfield, Illinois.
42. Himwich, W. A., and Himwich, H. E. (1959). Neurochemistry of aging. *In* "Handbook of Aging and the Individual, Psychological and Biological Aspects," (J. E. Birren, ed.), pp. 187-215. Univ. of Chicago Press, Chicago.

XII

Physiology of Work at Altitude

S. M. HORVATH

Introduction

Attempts to delineate the influence of high altitude residence (transient or permanent) on the working capacity and performance of man have been relatively inadequate, probably since all the factors that determine man's performance have not been identified and understood, even under the more optimal conditions for studying man at sea level. A complex multiple of physiological factors varying from respiratory and cardiovascular to tissue components are involved in determining the ability of man to perform work. Additional conditions making the analysis difficult are type and duration of exercise, ambient temperature, age, sex, initial states of physical condition, body composition, nutritional state, degree of activity while at altitude, and the level of altitude acclimatization attained by subjects. Further complications in analysis of published data relate to the wide variation in individual response to the combined stresses of low ambient oxygen pressure and level of work. It is also apparent that certain cultural and behavioral conditions modify the responses reported. Some subjects studied at altitude have resisted undertaking any physical activity except the tests planned, while others have engaged in additional and, to some extent, uncontrolled physical activity. Dr. Dill has conducted studies on man at altitude since 1929, and in his summary of the work carried out by the Fatigue Laboratory in his book, "Life, Heat and Altitude" (1), he has considered most of the points mentioned above. It is a tribute to his capabilities as an investigator in this field that most of the studies to be discussed reflect not only his early, original contributions but the continued impact he has had on investigations in altitude physiology up to the present. He is planning another study at the White Mountain laboratories after the celebration of his eightieth birthday.

Effects of Altitude on Work Performance

One of the first comprehensive studies on maximum oxygen uptake at altitude was made by Christensen in 1937 (2). In successive 10-day stays at four altitudes his maximum oxygen uptake of 3.72 liters at sea level showed a progressive decrease to 81, 69, 59, and 48% of his maximum at altitudes of P_B of 543, 489, 429, and 401 mm Hg, respectively. Studies conducted earlier at P_B 525 mm Hg had also shown a reduction in $\dot{V}_{O_2}$ from 78 to 92% of sea level values (1). The smallest reduction was associated with the longest period at this altitude. Buskirk (3) concluded that in unconditioned man $\dot{V}_{O_2}$ maximum was reduced 3.2% for each 1000 ft increase in altitude. The decrease was less for physically conditioned men, being approximately 2% for each 1000 ft. Apparently the threshold for this decrement was approximately 5000 and 20,000 ft. We have plotted the decreases reported in the literature (Fig. 1) and noted an exponential decrease. The only aberrant point was that from Pugh (4). According to some investigators, the reduction in $\dot{V}_{O_2 \max}$ at altitude improves with increasing duration of the sojourn. Dill *et al.* (5) and Klausen *et al.* (6) have reported gradual improvements in $V_{O_2 \max}$. Grover and Reeves (7) found in their studies on young runners that the $\dot{V}_{O_2 \max}$ was reduced 25% at 3100 meters. However, they also reported no improvement during 18 days at altitude nor on their return to sea level. Dill and Adams (8) also studied young runners at 3090 m and found a mean loss in $\dot{V}_{O_2 \max}$ of about 20% during the first week with a gradual improvement during the next 10 days.

The capacity for expending energy anaerobically at altitude decreased to a greater extent than the capacity for aerobic work. Edwards (9) had shown that the higher the altitude, the smaller the increase in blood lactate in all-out exercise. These findings have been frequently confirmed by others, the latest being the reports by Klausen (6) and Dill and Adams (8). This decrease in maximal blood lactate concentration continued throughout acclimatization of at least 6 weeks. It may be related to the reduction of alkali content of the blood at altitude, or it may be due to a reduction in glycogen stores in muscle, which is known to result in a reduced maximum lactate after maximal exercise.

Several investigators (6, 8, 10) have reported an increased $\dot{V}_{O_2 \max}$ of up to 10% following return to sea level in subjects who had been in residence for

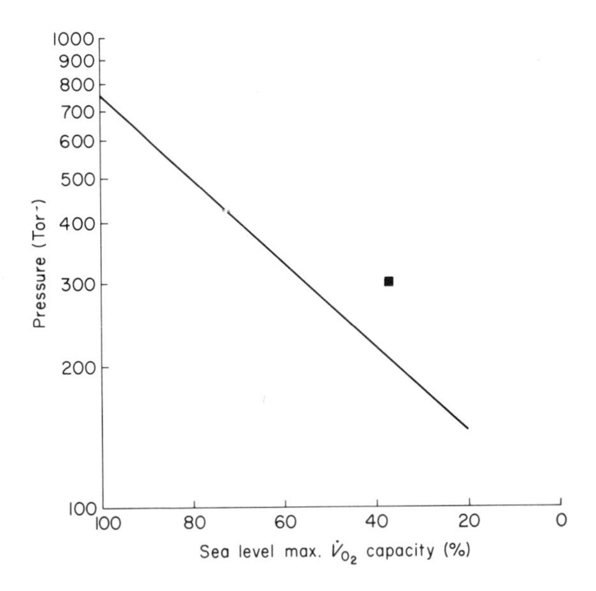

FIG. 1. Decrement in maximum $\dot{V}_{O_2}$ *capacity related to barometric pressure for transient residents at altitude.*

variable periods of time at altitude. Some of this improvement post exposure may be related to a general elevation in physical condition due to more frequent testing periods and overall increase in activity, or may be a reflection of training in anoxic states on metabolic competence at the tissue level. Grover and Reeves (7) and Saltin *et al.* (11) failed to find such a secondary rise in their highly trained runners. However, Dill and Adams (8), as well as Balke *et al.* (10, 12) reported that in their champion runners a small but significant increase in $\dot{V}_{O_2 max}$ occurred postaltitude. Hansen *et al.* (13) found no change in $\dot{V}_{O_2 max}$ in young men acclimatized at Pike's Peak.

The differences in work capacity between residents and transient sojourners at altitude have been reviewed by Hurtado (14) and Dill (1, 15), and there appears to be no question but that a lifetime of acclimatization has enabled natives to attain sea level performances. The natives had greater endurance and were superior in aerobic capacity. Workers at high altitudes appeared to be capable of doing heavy work without difficulty. Whether or not adult newcomers to altitude can ever become as effective as natives remains in question. Some investigators have implied that such a complete adaptation would be impossible, while others have suggested that many years of residence would be required to achieve similar levels.

A difference of opinion exists as to whether the efficiency with which work is performed at high altitude is altered. We have calculated the net mechanical efficiency of subjects who were at altitude for 12 days (Fig. 2). In young and middle-aged subjects there was a decreased efficiency, whereas in the oldest group (51-77 years of age) efficiency showed no change or a slight improvement. This latter difference was undoubtedly related to a training effect as was suggested by the measurements made on D. B. Dill (Fig. 3), who has been very active all of his life and exhibited a similar decrement in efficiency as seen in the younger subjects. Following return to sea level, all of the 16 subjects had improved their mechanical efficiency over their control levels. The significance of this increase remains unanswered but may be related to the higher $\dot{V}_{O_2 max}$ mentioned earlier and improved physical fitness.

The cardiovascular system makes early adjustments to high altitude (16). A given submaximal work load is accomplished with greater ease as the individual becomes acclimatized. Although there remain some questions as to the precise changes that occur during rest and moderate work at altitude, cardiac output, stroke volume, and heart rate generally exhibited increases during the first days, but gradually decreased to sea level values or slightly below as residence continued (17, 18). It is interesting that Hartley *et al.* (19) have found subnormal cardiac outputs and stroke volumes in residents at 3100 meters altitude. During severe (maximal) exercise the maximal heart rate attained was generally always less by some 20 beats/min. The magnitude of this decrease appeared to be dependent upon the state of physical fitness of the subject as

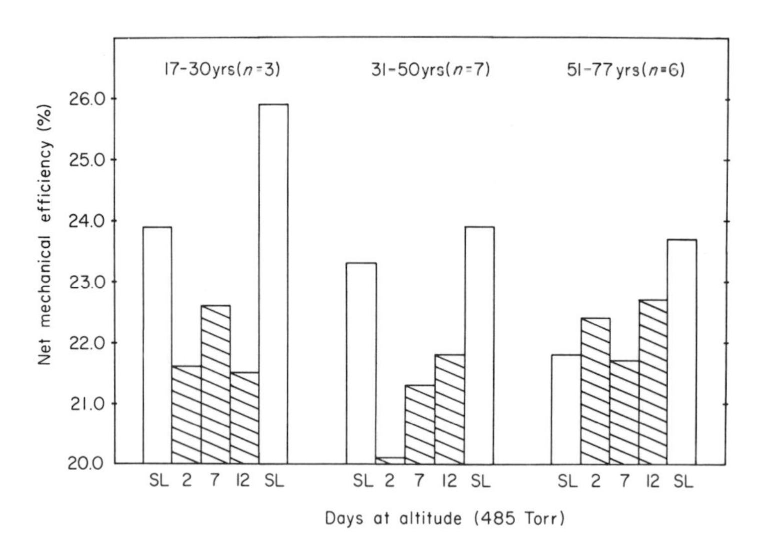

FIG. 2. Changes in mechanical efficiency during moderate work of 20-min duration before, during, and after (2 days) a period of residence at altitude as further influenced by the age of the subjects.

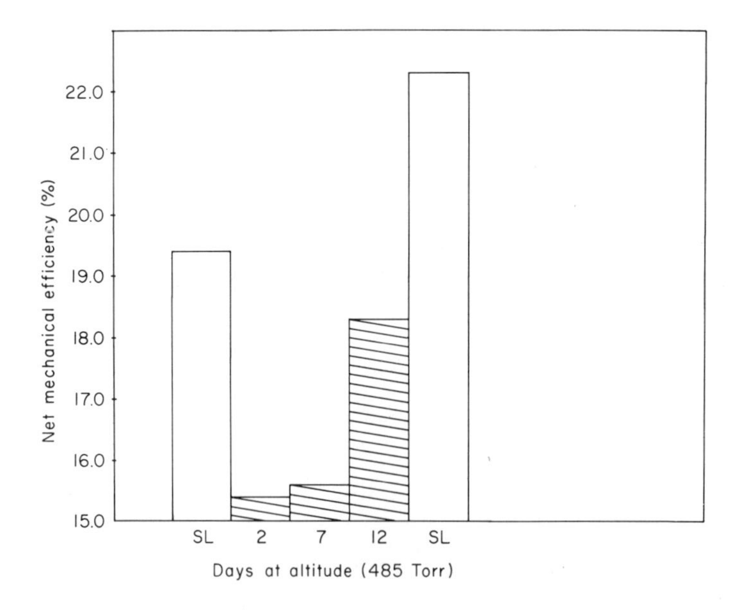

FIG. 3. Alterations in mechanical efficiency of the oldest subject (DBD, 77 years) studied during moderate work at altitude.

well as his age. Younger and more fit men had smaller decrements, whereas older subjects showed the greatest decrease. The inability at altitude to raise the cardiac output to the maximal levels observed at sea level remains unexplained (4, 20). It has been suggested that the increased hematocrit and the consequent higher viscosity of the blood was the major factor involved, but no positive evidence for this effect has been reported. Grover *et al.* (21) measured coronary blood flow at rest and at light work in subjects after 10 days at 3100 m and found a marked decrease. Apparently the increased requirements for oxygen by the myocardium were met mainly by an increased oxygen extraction and partially by the increased concentration of 2,3 diphosphoglycerate in the red blood cells. These findings suggest that the 2,3 DPG influence on shifting the oxygen dissociation curve to the right, decreasing the affinity of hemoglobin for oxygen, may alleviate the lack of coronary vasodilation and so maintain myocardial oxygen tension at optimal levels. Consequently, the inability of man to elevate his cardiac output may be limited simply by the failure of the coronary blood flow to reach the levels required for this amount of cardiac work.

Although many investigators have studied the response of man to maximal as well as to submaximal work loads, only a few have directed their attention to sustained submaximal levels. Billings *et al.* (22) determined the oxygen uptake of men living at 3800 m and working for 30 min at 50% and 67% of their sea level maximum capacity. They reported that work which required an oxygen uptake of 2.2 liters or less could be sustained at about the same oxygen cost at altitude as at sea level. Above this level of work, i.e., at oxygen uptakes in excess of 2.5 liters, a lower oxygen uptake was observed at the higher altitude. These findings suggested that a larger oxygen debt was accumulated as a result of this higher sustained level of work. All subjects could not maintain this level of work for the required 30 min, even after 3 weeks of acclimatization to altitude. Reeves and Daoud (23) have presented some preliminary results of their studies on long-term work of 2 hrs duration at an altitude of 15,000 ft (P_B = 425 mm Hg). The subjects were acutely exposed to this altitude. Although all six subjects were able to complete their 2 hours of work, most required encouragement to finish the task. All became unsteady and found it necessary to hold the hand rail of the treadmill. During the control walk at 1000 ft at zero grade and 3 mph, $\dot{V}_{E_{BTPS}}$ was 27.7 liters, oxygen uptake was 1.14 liters, and heart rate was 98 beats/min. At a similar work load beginning shortly after reaching altitude these values were, respectively, 41.6, 1.34, and 130. These findings strongly suggest the need for additional studies on long-term work at altitude.

Dill *et al.* (5) measured the work capacity at sea level and at altitude, P_B 455 mm Hg of five men who had taken part in similar studies at high altitude from 18 to 33 years earlier. The maximum oxygen uptake declined with age both at sea level and at altitude. There was considerable difference in the response of each individual in the rate of his adaptation. One subject, JT, at age 26 was able to attain a level of 56% of his maximum, while at age 60 a value of 48% was found. On the other hand, Dill, the oldest subject at 71 years, reached 76% after 1 week at altitude but had increased his capacity to 86% after 5 weeks of acclimatization. The responses in general were similar to those seen in younger subjects, namely an initial decrease during the first days at altitude followed by progressive improvement with continued stay at altitude. Whether or not older individuals can become fully acclimatized to altitude remains an unanswered question. Since significant individual differences in physiological capabilities for adaptation were observed, and, in view of the occurrence of different mechanisms of adaptation in older subjects such as the failure of older subjects to exhibit the hemoconcentration and decrease in plasma volume seen in young men in the early days of altitude exposure, it may be that the adaptative mechanisms for exercise are also different. However, it was quite apparent that although the pattern of improvement in maximal physical work capacity was similar at all ages, the rate of such improvement was different in the two groups. Further investigation of the age-related capacities and physiological adaptations is imperative to an understanding of the processes of adaptation.

A number of ergogenic aids have been utilized by investigators in attempts to improve altitude tolerance and work performance at altitude. The substances utilized either altered the acid-base balance, removed fluid from the pulmonary vascular bed, improved oxygen transport, and/or modified the response of the cardiovascular system. Methylene blue was one of the first agents investigated but proved to be ineffective and toxic. Cholinergic drugs were said to improve tolerance, whereas adrenergic drugs reduced resistance. Ammonium chloride and potassium salts were also employed with variable degree of effectiveness. Carbonic anhydrase inhibitors have been frequently utilized in recent years. A diuretic, furosemide, has been recently reported to improve physical work capacity. Nair and Gopinath (24) gave furosemide to young men who had been acclimatized to 11,000 ft for 21 days. They reported an increased oxygen extraction, higher oxygen uptake, and a decreased oxygen debt with no change in pulmonary ventilation as a consequence of a single dose of this substance. The implication from these studies, that there was a reduction in pulmonary blood volume and consequently an improvement in the transfer of oxygen from the alveoli to the pulmonary capillaries, was not tested by measurements of pulmonary blood volumes. It is evident that further and more complete investigation of the potential benefit of ergogenic aids must be undertaken.

The lack of agreement as to the degree of impairment, potentialities for improved performance, and the time sequence of change in work capacity at various altitudes suggests the need for additional studies, despite the extensive investigations already completed. There is a further need to consider levels of exercise other than maximum capacity tests, to relate these to the percentage of maximum performance, and finally to consider the physiological alterations during and consequent to extended periods of physical activity requiring hours and perhaps days of such work loads.

References

1. Dill, D. B. (1938). "Life, Heat and Altitude," p. 211. Harvard Univ. Press, Cambridge, Massachusetts.
2. Christensen, E. H. (1937). Sauerstoffaufnahme und respiratorische Functionen in grossen Höhen. *Skand. Arch. Physiol.* **76**, 88.
3. Buskirk, E. R. (1969). Decrease in physical working capacity at altitude. *In* "Biomedicine Problems at High Terrestrial Elevations" (A. H. Hegnauer, ed.), p. 323, Natick, Massachusetts U.S. Army Institute of Environmental Medicine has concluded that in unconditioned men $\dot{V}_{O_2}$ max is reduced 3.2% for each 1000-ft increase in altitude and only 1.9% for conditioned men; apparently the threshold is somehwere around 5000 ft, whereas from 5000 to 22,000 ft the decrease is linear.
4. Pugh, L. G. C. E. (1964). Animals in high altitudes: Man above 5,000 meters–mountain exploration. *In* "Handbook of Physiology" (Amer. Physiol. Soc., J. Field, ed.), Sect. 4, p. 861. Williams & Wilkins, Baltimore, Maryland.
5. Dill, D. B., Robinson, S., Balke, B. and Newton, J. L. (1964). Work tolerance: Age and altitude. *J. Appl. Physiol.* **19**, 483.
6. Klausen, K., Dill, D. B., and Horvath, S. M. (1970). Exercise at ambient and high

oxygen pressure at high altitude and at sea level. *J. Appl. Physiol.* **29**, 456.

7. Grover, R. F., and Reeves, L. T. (1967). Exercise performance of athletes at sea level and 3,100 meters altitude. *In* "The Effects of Altitude on Physical Performance" (R. F. Goddard, Ed.), p. 177. Athletic Institute, Chicago, Illinois.
8. Dill, D. B., and Adams, W. C. (1971). Maximal oxygen uptake at sea level and at 3,090 m altitude in high school champion runners. *J. Appl. Physiol.* **30**, 854.
9. Edwards, H. T. (1936). Lactic acid in rest and work at high altitudes. *Amer. J. Physiol.* **116**, 367.
10. Balke, B., Faulkner, J. A., and Daniels, J. T. (1966). Maximum performance capacity at sea level and at moderate altitude before and after training at altitude. *Schweiz. Z. Sports Med.* **14**, 106.
11. Saltin, B., Grover, R. F., Blomqvist, C. G., Hartley, L. H., and Johnson, R. L., Jr. (1968). Maximal oxygen uptake and cardiac output after two weeks at 4,300 meters. *J. Appl. Physiol.* **25**, 400-409.
12. Balke, B., Nagle, F. J., and Daniels, J. (1965). Altitude and maximal performance. *J. Amer. Med. Ass.* **194**, 646.
13. Hansen, J. E., Vogel, J. A., Stelter, G. P., and Consolazio, C. F. (1967). Oxygen uptake in man during exhaustive work at sea level and high altitude. *J. Appl. Physiol.* **23**, 511.
14. Hurtado, A. (1964). Animals in high altitudes: Resident man. *In* "Handbook of Physiology" (Amer. Physiol. Soc., D. B. Dill, ed.), Sect. 4, p. 843. Williams & Wilkins, Baltimore, Maryland.
15. Dill, D. B. (1968). Physiological adjustments to altitude changes. *J. Amer. Med. Ass.* **205**, 747.
16. Horvath, S. M., and Howell, C. D. (1964). Organ systems in adaptation: The cardiovascular system. *In* "Handbook of Physiology" (Amer. Physiol. Soc., D. B. Dill, ed.), Sect. 4, p. 153. Williams & Wilkins, Baltimore, Maryland.
17. Klausen, K. (1966). Cardiac output in man in rest and during work and after acclimatization to 3,800 m. *J. Appl. Physiol.* **21**, 609.
18. Saltin, B. (1966). Aerobic and anaerobic work capacity at an altitude of 2,250 meters. *In* "The International Symposium on the Effect of Altitude on Physical Performance," (R. F. Goddard, ed.), pp. 97-102. Athletic Institute, Chicago, Illinois.
19. Hartley, L. H., Alexander, J. K., Modelski, M., and Grover, R. F. (1967). Subnormal cardiac output at rest and during exercise in residents at 3,100 m. *J. Appl. Physiol.* **20**, 839.
20. Pugh, L. G. C. E. (1964). Cardiac output in muscular exercise at 5,800 m. *J. Appl. Physiol.* **19**, 441.
21. Grover, R. F., Lufschanowski, R., and Alexander, J. K. (1970). Decreased coronary blood flow in man following ascent to high altitude. *In* "Hypoxia, High Altitude and the Heart" (J. H. K. Vogel, ed.), p. 72. Karger, Basel.
22. Billings, C. E., Bason, C. R., Mathews, D. K., and Fox, E. L. (1971). Cost of submaximal and maximal work during chronic exposure to 3,800 m. *J. Appl. Physiol.* **30**, 406-408.
23. Reeves, J. T., and Daoud, F. (1970). Increased alveolar-arterial gradients during treadmill walking at simulated high altitudes. *In* "Hypoxia, High Altitude and the Heart" (J. H. K. Vogel, ed.), p. 41. Karger, Basel.
24. Nair, C. S., and Gopinath, P. M. (1971). Effect of furosemide (Lasix) on physical work capacity of altitude-acclimatized subjects at an altitude of 11,000 feet. *Aerosp. Med.* **42**, 268.

XIII

The Hypoxias of Altitude, of Anemia, and of Carbon Monoxide Poisoning

W. H. FORBES

The oxygen supply carried by the blood to the tissues has two main characteristics that affect the performance of these tissues–the content and the pressure of O_2. These two variables are obviously related. If there is a high content of O_2 in the blood the average pressure in the tissues will be higher than if there is a low content, as in anemia, even though the initial pressure in the anemic blood may be equal to that in the normal blood. By varying other parameters, particularly the amount of hemoglobin available or the rate of circulation, the content and the pressure can be varied separately, even though they are not independent. One can thus get four markedly different situations–both content and pressure high; both content and pressure low; content high and pressure low; content low and pressure high. These are more clearly kept in mind if the circumstances with which they are associated are considered. In the following discussion sea level pressures of O_2, i.e., 155-160 mm Hg, will be considered "high" as they are the highest ordinarily encountered in nature, and anything under 90 mm Hg of O_2 in the inspired air (= 14,000 ft) is considered low. This figure is chosen because this is about as high as people live in any number, although there are small groups living up to 17,000 ft or even higher.

In polycythemia at sea level, e.g., an Andean mountaineer descending to the lowlands, both content and pressure are high. In an anemic person going up into the mountains, both are low. In the third case one must go to a high-altitude animal, such as a llama, which will have a high content even when the ambient pressures are relatively low; finally, in the fourth case, an anemic man at sea level will have a low content at a high pressure. The difficulty of finding a man to fill the third situation—a high content at a low pressure—is due to the position of the human hemoglobin dissociation curve (too far to the right). At low pressures, i.e., at an altitude 17,000 ft or more the saturation is below 80% and falls rapidly at higher altitudes, and the polycythemia produced by prolonged residence at altitudes does not keep up with the fall in the percentage saturation, so the cubic centimeters of oxygen carried per 100 cc of blood falls slowly with altitude. However a man acclimatized at 18,000 ft, who descends to about 10,000 ft, will have a high content at a moderate pressure, though he will not have a high content at a low pressure.

The easiest and commonest way of producing the effects of altitude at sea level in the absence of a low-pressure chamber is by breathing mixtures of air and nitrogen. This works very well until one gets to very high rates of ventilation as in strenuous exercise. At high rates the greater density of gas at 760 mm Hg and its frictional resistance begin to be significant in increasing the work of breathing.

The effects of anemia if it is produced suddenly, as in profuse bleeding, are complicated by the effects of lowered blood pressure and are characterized by fainting rather than by the mental confusion and headache of altitude or carbon monoxide poisoning. If the anemia comes on slowly, as in dietary iron deficiency, there is an extraordinary amount of adaptation to it especially in children who can tolerate hemoglobins down to 4 gm% or even a trifle lower, which is one-fourth of the normal amount. Adults also adapt surprisingly well, being able to walk about (slowly) with hemoglobins of 4 gm. In these cases inactivity and a greatly increased circulation are the main adaptations, and a significant fraction of the oxygen is carried in solution in the plasma. Anemia and carbon monoxide poisoning are analogous except in their rate of onset and in the viscosity of the blood. As anemias in man are less suitable for experimentation than carbon monoxide poisoning they will not be further discussed.

Carbon monoxide poisoning is another condition in which the effects appear to be wholly due to anoxia and therefore comparable to altitude. Originally carbon monoxide was considered an exogenous poison and suspected of having poisonous effects in addition to its displacement of oxygen, but the work of Sjöstrand (1) demonstrated that carbon monoxide was constantly produced in the body in the course of the normal breakdown of hemoglobin, and this has changed the thinking about its effects. It now seems almost certain that its only deleterious action is the displacement of O_2 from hemoglobin molecules. However there is not complete agreement on this point (2).

There is one important difference between the anoxia of carbon monoxide poisoning and that of altitude or of air and nitrogen mixtures. In carbon monoxide poisoning the amount of O_2 carried by the blood is decreased but its pressure is only very slightly changed. This slight change is due to a shift to the left in the O_2 dissociation curve. As some of the hemoglobin combines with carbon monoxide the remaining O_2 Hb retains its oxygen a little more strongly and therefore slightly lowers the O_2 tension in the blood. A 10% saturation of the blood with carbon monoxide would therefore be expected to have a slightly greater adverse effect on the O_2 supply than would withdrawing 10% of the cells and replacing them with plasma.

However, if the 10% diminution in the O_2 Hb were produced by a lowering of the O_2 tension in the lung, this lowering would have to be from 100 mm Hg to about 65 mm Hg. As a result one would expect an effect on the organism at least as great and probably definitely greater than that produced by 10% CO Hb. This expectation is supported by impressions from experimental observations. At relatively low CO Hb levels (highest 26%) Dr. von Post-Lingen (2) in speaking of her experiments on carbon monoxide poisoning says "The plan was to observe the reactions of healthy persons to CO in concentrations which do not generally cause subjective disturbance, i.e., which produce carboxyhaemoglobin (CO Hb) below 25%." Dr. von Post-Lingen found that 38 out of 100 subjects who reached 16-26% CO Hb reported dizziness, fatigue, and headache some hours after inhaling the carbon monoxide. In contrast to this almost all unacclimatized persons going to Morococha (14,800 ft) have mountain sickness even though there is usually a period of 5 hr or so during which they ascend gradually. At Morococha the arterial saturation is around 80%. Unfortunately there is a flaw in this comparison as the exposures to carbon monoxide in Dr. von Post-Lingen's studies, even though abrupt in onset, were presumably shorter in duration than the exposures at Morococha. The times are not given in Dr. von Post-Lingen's article.

As the severity of the exposure is increased the lesser effect of carbon monoxide poisoning becomes more obvious. With 40% CO Hb and the rest in equilibrium with air at sea level (i.e., nearly 60% O_2 Hb) there is headache, weakness, a tendency to faint if muscular work is attempted, and generally unpleasant feelings, but there is no danger to life (3). The sudden exposure to an altitude of 21,500 ft, which reduces the arterial saturation to 60%, is likely to cause loss of consciousness within 10 min (4) and death in a few hours in an unacclimatized person. The effects of reducing the O_2 Hb to 55% or 50% (3) are even more dramatic. If this reduction is produced by carbon monoxide, simple tasks (simulated driving an automobile) can still be done if no muscular work is required (3) but are done poorly. If the O_2 Hb is lowered to 50% by exposure to altitudes (24,000 ft) consciousness may be lost in 3 or 4 min.

Most physiologists have held the view that because the oxygen tensions are higher in carbon monoxide poisoning than they are with an equivalent lowering of the O_2Hb from altitude, the latter is the more severe strain upon the organism as indeed it is when 50% O_2 Hb is approached. It has always been commonly assumed that this is also true at a lower altitude and at a lower concentration of carbon monoxide. Recently at the Chicago meetings a paper (5) was presented in which the effects of levels of 80% arterial saturation with O_2 produced by these two methods were compared and found to be equal. The subjects were doing heavy work which introduces a factor not found in many comparative experiments on carbon monoxide poisoning, but there seems to be no obvious reason why this should make a difference.

At first glance (and there has hardly been time for more) the slight shift to the left in the O_2 dissociation curve when carbon monoxide is present seems a possible explanation of the lack of the expected difference between the effects of carbon monoxide and of low oxygen. Indeed it is the most attractive explanation, but it hardly seems adequate. The effect of 20% CO Hb on moving the O_2 dissociation curve to the left is small, a matter of 2 or 3 mm of pressure, but the difference in O_2 pressure in the lungs and therefore in the plasma between sea level ($O_2 = 100 \pm$ mm) and 14,000 ft altitude (O_2 Hb = 80%, $p_{O_2} = 55$ mm) is about 45 mm. Another factor which would tend to diminish the difference, but not eliminate it, is the fact that the men were working hard and therefore a greater proportion of the O_2 was carried by the cells and a smaller proportion by the plasma than would be the case in rest. Finally, there is the possibility that carbon monoxide interfered in some slight way with the transfer of oxygen to the myoglobin. In any event these observations suggest further studies.

References

1. Sjöstrand, T. (1951). Endogenous formation of carbon monoxide: The CO concentration in the inspired and expired air of hospital patients. *Acta Physiol. Scand.* **22**, 137.
2. von Post-Lingen, M. L. (1964). The significance of exposure to small concentration of carbon monoxide. Experimental study on healthy persons. *Proc. Roy. Soc. Med.* **57**, Part 2, 1021.
3. Forbes, W. H., Dill, D. B., DeSilva, H., and Van Deventer, F. M. (1937). The influence of moderate carbon monoxide poisoning upon the ability to drive automobiles. *J. Ind. Hyg. Toxicol.* **19**, 598.
4. McFarland, R. A. (1953). "Human Factors in Air Transportation," p. 177. McGraw-Hill, New York.
5. Vogel, J. A., Gleser, M. A., and Mello, R. P. (1971). Oxygen transport during exercise with carbon monoxide exposure. *Fed. Proc., Fed. Amer. Soc. Exp. Biol.* **30**, 371.

XIV

Physiology of Respiration at Altitude

B. BALKE

The physiological effects of altitude on respiration provide a never ending source of potential research problems. No other environmental stress allows us to study in greater detail the adaptability and limitations of organs and organ systems to the variety of metabolic loads from rest to maximal physical exertion.

It is beyond the scope of this chapter to present a review of the research accomplished during the last 40 years by respiratory physiologists of name and rank.

Dr. Luft (Chapter X) on principles of adaptation to altitude has pointed out the facts that affect respiratory functions at altitude without alluding to them in detail. I will restrict myself to pointing out some problems that still exist with regard to the work and regulation of pulmonary ventilation in acute and chronic hypoxia.

Teleologically, at first thought, it would seem desirable that under the influence of deprived oxygen in the atmospheric air of high altitude the need for oxygen should be reduced as a most meaningful adaptation. On second thought, however, that would not make too much sense because even at lower demands the supply would eventually become insufficient.

The older as well as the newer literature is rather controversial on this point. Careful studies of the basal metabolic rate by Douglas *et al.* (1) in 1912 and by Barcroft (2) in 1927 at an altitude of 4500 m, by Herxheimer, Kost, and Ryjaczek (3) in 1933 at 3470 m, and by Balke (4) at elevations of 3000 and 3450 m showed a 0, 2.5, 2 and 1.4% increase over controls, respectively. Other studies (5-9) revealed increases from as low as 3.8% at an elevation of 2130 m (10) to 49% at 4500 m (5). Jaquet (11) reported an increase of 8.8% for especially "altitude sensitive" individuals at an elevation as low as 1600 m. Grover (12), who obtained control data at about this elevation and altitude data at 4320 m concluded that when the resident of low altitude is exposed to high altitude, his basal oxygen uptake remains unchanged or increases slightly. Dill (13) who had seen no change from control values in 2 individuals at 3100 m in 1929 reported an increase from 10 to 13% in the same two and other 4 subjects 34 years later at nearly the same elevation.

The only real *reduction* of the resting metabolic rate at altitude was reported for steers and lambs (14, 15). In the case of the latter, a decrease in body temperature of 2°C could have lead to the decrease in metabolism. The steers, it was hypothesized, might have been in a more basal state.

In a study of the supine resting metabolic rate of people undergoing physical training at low and relatively high elevations in the Alps, I have made the observation that on the morning after a day of strenuous mountaineering at altitudes around 2000 m the "basal" metabolic rate of the well-trained individual was normal but that of the less trained was elevated. On the other hand, at the higher elevations above 3000 m a similarly stressful day of mountain climbing resulted in increased resting metabolic rates for 24-36 hr in the well-trained individuals as well. This was hardly explainable as a consequence of still elevated body temperature because the usually cold environmental conditions were more likely inducive to relatively low core temperatures. Since thermoregulatory muscle tension or actual shivering can affect total energy expenditure, care was taken to keep the experimental subjects as warm as possible. Another possibility of measuring different levels of resting oxygen consumption even in the fasting state is the diurnal variability of the BMR, as was shown in Chapter VII. There are usually two metabolic peaks, one at about noon and the other in the evening hours, in spite of restricted food intake and activity levels. For that reason, only values obtained during the same hour of the day should be used for comparison.

During moderate to severe exercise, i.e., under conditions that are usually more inducive to precise physiological regulations than the resting state, oxygen costs for identical work intensities remain the same at different levels of altitude (3, 16-18).

As an illustration, Fig. 1 may show that this was true for a group of well-trained people of nearly identical aerobic work tolerance levels. A standard test procedure on the bicycle ergometer was applied first under sea level conditions, then at the simulated altitude of 4260 m in a low pressure chamber after 3-4 hr of exposure to attain an "equilibrium," and somewhat later at the

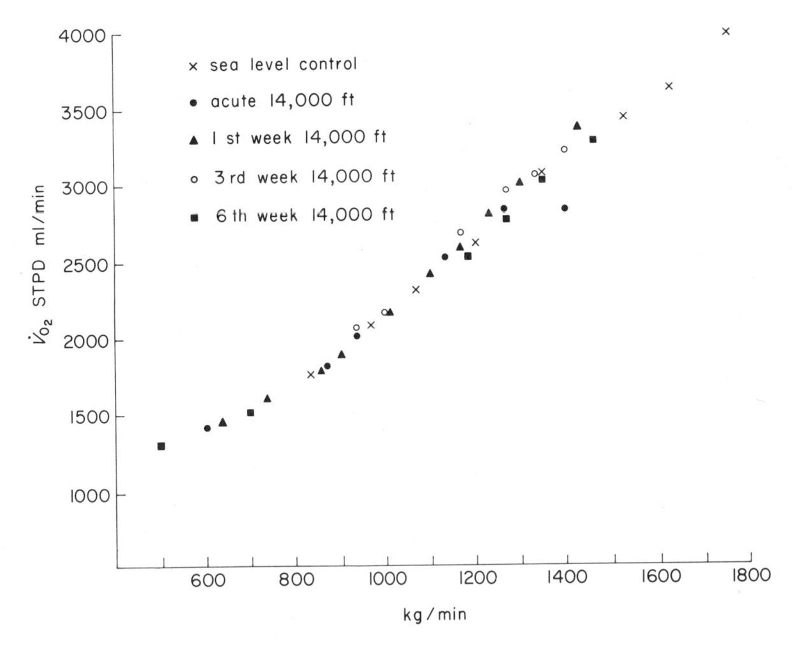

FIG. 1. Relation of fixed work loads on the bicycle ergometer to oxygen consumption at sea level and during the course of acclimatization to an altitude of 14,000 ft (4260 m).

true elevation of 4260 m during the first, third, and sixth week of acclimatization. Regular physical training was continued at altitude.

There was no significant change in the $\dot{V}_{O_2}$ for any given work intensity under these various experimental conditions.

In another series of experiments (19), animals did not behave differently than man. Out of several dogs trained to run on the treadmill and to perform a standardized treadmill test similar to the one applied to man, four dogs were kept for 5 months in a low pressure chamber at a simulated altitude of 5780 m. After this exposure they were retested twice on the treadmill under sea level conditions, once breathing ambient air, the other time an oxygen-nitrogen mixture equivalent to an altitude of 5780 m. Although maximum aerobic work tolerance was different under the three different test conditions (pre and post acclimatization sea level controls, post acclimatization hypoxia test run), oxygen consumptions for given levels of work were practically the same.

These experimental observations have been reemphasized to point deliberately to the varying demands on pulmonary ventilation at different altitudes at which the molecular oxygen requirements for given energy expenditures during rest or work remain constant. Values of oxygen consumed or of carbon dioxide exhaled are, as a rule, expressed under conditions standardized with regard to atmospheric pressure, temperature, and water vapor content. However, the "ideal" condition of P_B = 760 mm Hg, T = 0°C, and P_{H_2O} = 0, does not exist in the lungs where the volume of air actually moved to correspond with a given $\dot{V}_{STPD}$ at various altitudes must change to compensate for the decrease or increase of air density. Man's respiratory pump can be compared to any mechanical pump that delivers flow at a magnitude depending on bore, stroke, and frequency. In contrast to most pumps or piston engines, in which bore and stroke remain constant and only stroke frequency determines the total output in time, the ventilatory pump of man can adjust to increasing or decreasing demands not only with changes of the respiratory frequency but with changes of "bore and stroke" as well.

Thus, a decrease in air density at high altitude could be compensated for by (a) an increase of the bore-stroke-frequency product, i.e., by augmented mechanical bellow action or (b) shifts in the pulmonary gas concentrations toward lower alveolar oxygen and higher carbon dioxide fractions. The latter type of compensation has been well established in acute altitude exposures by Boothby *et al.* (20). However, the conditions of "homeostasis" in the lungs are then upset and such compensation can only be of temporary nature.

It has been pointed out by Rahn and Otis (21) and by Riley *et al.* (22) that during acute exposure to altitudes lower than 3500 m or to alveolar oxygen tensions above 60 mm Hg subjects generally show no increase in pulmonary ventilation (BTPS). Because of the flat part of the oxygen dissociation curve downward from 100 to 60 mm Hg the oxygen saturation of the arterial blood is not markedly influenced, and the hypoxic respiratory drive is nearly negligible.

The iso-$\dot{V}_{BTPS}$ at sea level and in acute exposure to altitude constitutes a hypoventilation with regard to the $\dot{V}_{STPD}$ at altitude, resulting in practically the same alveolar CO_2 tension as at sea level. One could conclude therefore that the alveolar or arterial CO_2 pressure remains the major respiratory drive in acute altitude exposure. With the decrease in atmospheric pressure the partial pressures of all alveolar gases are lowered. The reduced $P_{A_{CO_2}}$ then inhibits the normal stimulation of the respiratory center, causing the relative hypoventilation until the $P_{A_{CO_2}}$ has risen to the level normally required to provide the appropriate ventilatory stimulus.

Since pulmonary ventilation is mainly geared to the metabolic rate and probably more closely to the elimination of carbon dioxide than to the demand for oxygen—at least at more normal oxygen pressure—a closer look at the changes of gas concentrations during the respiratory cycle might be worthwhile. Possibly, the magnitude of change in $P_{A_{CO_2}}$ may help to explain the maintenance of rhythmic ventilation at all metabolic rates.

The following assumptions are made as a basis for further computations:

Resting conditions, at $P_B = 747$ mm Hg

$\dot{V}_{O_2} = 240$ ml/min	$\dot{V}_{CO_2} = 200$ ml/min	
$\dot{V}_{BTPS} = 6.2$ liters/min	$\dot{V}_{STPD} = 5.0$ liters/min	$f = 12$
$F_{E_{CO_2}} = 0.04$	$F_{A_{CO_2}} = 0.057$	$P_{A_{CO_2}} = 39.9$ mm Hg
$F_{E_{O_2}} = 0.163$	$F_{A_{O_2}} = 0.143$	$P_{A_{O_2}} = 100$ mm Hg
$FRV_{BTPS} = 3500$ ml	$FRV_{STPD} = 2820$ ml	
$V_{T_{BTPS}} = 516$ ml	$V_{D_{BTPS}} = 186$ ml	$V_{(T-D)STPD} = 266$ ml

Furthermore, we will assume that at the respiratory frequency of 12 breaths/min the inspiratory phase requires 2 sec and the expiratory phase 3 sec. Then the following calculations, presented in Table I, may show the possible changes of volume, gas content, and concentration during various fractions of the respiratory cycle for sea level conditions as well as for an altitude of 9000 ft, with the BTPS values of all ventilatory volumes supposedly remaining the same.

A comparison of the alveolar CO_2 tensions during the respiratory cycle shows that they are nearly the same at both altitudes. If iso-$\dot{V}_{STPD}$ were to persist at altitude, the $P_{A_{CO_2}}$ would be too low to serve respiratory stimulation.

The slight difference between the "measured" end-expiratory value at zero time and the calculated value at the end of the breathing cycle can be explained by slight time delays between the end of expiration and beginning of inspirations. During the resting state the additional accumulation of CO_2 has only little effect compared to that during exercise. Two factors contribute in exercise to a quicker and more drastic change of the CO_2 content in the alveolar air of the functional residual volume: the first is the increased metabolic production of CO_2; the second factor is the decrease of FRV as the tidal volume encroaches more and more, with increasing work intensities, on the expiratory reserve volume.

For the condition of exercise requiring 2 liters/min of oxygen consumption, the following values (all volumes at BTPS) might be assumed at sea level:

$\dot{V}_{CO_2} = 1820$ ml/min	$F_{E_{CO_2}} = 0.045$	$F_{A_{CO_2}} = 0.051$
$\dot{V} = 50$ liters/min	$V_T = 2500$ ml	$V_D = 300$ ml, $FRV = 2500$ ml

With a respiratory frequency of 20 breaths/min a breathing cycle lasts 3 sec, whereby it will be assumed that inspiration is completed within 1 sec, expiration in 2 sec—with two-thirds of $\dot{V}_T$ expired in the first, the other third in the second second.

The tabulation of calculated values presented in Table II represent the conditions for same work at sea level and in acute altitude exposure to ~ 3000 m. All volume expressed at STPD condition.

TABLE I Alv CO_2 Changes during Respiration Cycle at Rest

REST	P_B = 747 mm Hg					P_B = 547 mm Hg				
	0	2 sec	3 sec	4 sec	5 sec	0	2 sec	3 sec	4 sec	5 sec
$V_{(T-D)}/t$		266	−150	−100	−16		191	−108	−72	−11
FRV	2820	3086	2936	2836	2820	2030	2221	2113	2042	2030
F_{ACO_2}	0.057	0.0542	0.0554	0.0566	0.0578	0.0792	0.0754	0.0768	0.0784	0.08
$FRV \cdot F_{ACO_2}$	160.7	167.5	162.7	160.5	162.9	160.7	167.5	162.4	160.2	162.7
F_{ACO_2}		0.1					0.08			
$V_{(T-D)} F_{ACO_2}$			−8.13	−5.54	−0.89			−8.4	−5.53	−0.87
$\dot{V}_{CO_2}/t$		6.66	3.33	3.33	3.33		6.66	3.33	3.33	3.33
P_{ACO_2}	39.9	37.7	38.4	39.3	40.1	39.2	37.4	38.2	39.0	39.8
P_{ACO_2} for iso-$\dot{V}_{STPD}$ condition						28.5	27.2	27.7	28.3	28.9

TABLE II. Alv CO_2 Changes during Respiratory Cycle in Exercise

	P_B = 747 mm Hg				P_B = 547 mm Hg			
Exercise	*0*	*1 sec*	*2 sec*	*3 sec*	*0*	*1 sec*	*2 sec*	*3 sec*
V_{T-D}/t		1780	−1200	−580		1275	−850	−425
FRV	2020	3800	2600	2020	1450	2725	1875	1450
F_{ACO_2}	0.051	0.035	0.047	0.061	0.071	0.049	0.065	0.085
$FRV \cdot F_{ACO_2}$	103	134	121	124	102	133	121	124
$V_{T-D} \cdot F_{ACO_2}$		0.7				0.5		
F_{ACO_2}			−42.3	−27.1			−41.3	−27.5
VCO_2/t		30.0	30.0	30.0		30.0	30.0	30.0
P_{ACO_2}	35.7	24.6	32.7	43.0	35.3	24.4	32.3	42.9

The question that has not been answered at present is whether these alveolar and blood gas oscillations can elicit responses from the arterial chemoreceptors. Purves (23) has measured oscillations in P_{O_2} in the carotid artery. Dubois and Yamamoto (24) have shown that the numeral component of exercise hyperpnea was abolished when the arterial blood was stirred before it reached the arterial chemoreceptors.

In all probability, during exercise of high metabolic costs the changes of alveolar P_{CO_2} and P_{O_2} during the respiratory cycle are of greater importance for the regulation of respiration than presently realized.

In the resting state these fluctuations are relatively small, and ventilation is readily subjected to voluntary control. In exercise, once the return of metabolically produced CO_2 to a reduced end-expiratory residual volume is high, it is nearly impossible to hold the breath voluntarily or to extend the duration of expiration for only a very short time. If the breathing cycle in the exercise situation mentioned above would have been extended by just 0.5 sec, a P_{ACO_2} of about 48 mm Hg would have resulted. Although at sea level the corresponding value of alveolar oxygen tension would not have been low enough to affect oxygen saturation seriously, at a P_B of 547 mm Hg it would have been decreased to 60 mm Hg or below. At that P_{AO_2} the newcomer's hypoxic respiratory drive becomes effective at rest. The exercise situation is probably inducive to hyperventilatory adjustments in the newcomer to altitude. There seems to be a carry-over of the increased ventilatory response into the resting state, thus enhancing the respiratory adaptation to hypoxia. Ideally, optimal ventilatory conditions would be reestablished by bringing the $\dot{V}_{STPD}$ back to former sea level values. Experiments, illustrated in Fig. 2, have shown that, in acute exposure to a simulated altitude of 4260 m in a low pressure chamber, $\dot{V}_{STPD}$ was lower than at ground level and remained so after 1 week and even 3 weeks of altitude acclimatization in the mountains. Only after 6 weeks of acclimatization to 4260 m, possibly enhanced by physical training, did $\dot{V}_{STPD}$ return to sea level values.

The same trend was observed when relating $\dot{V}$ to $\dot{V}_{CO_2}$ (Fig. 3), namely a relative hypoventilation during the acute exposure, which disappeared with progressing acclimatization.

This "hypoventilation" of the newcomer was not only seen during exercise but also at rest in experiments with progressing hypoxia as produced by lowering the atmospheric pressure in the chamber. Figure 4 shows that pulmonary ventilation remained below the theoretical values estimated for a given $\dot{V}_{O_2}$ and $\dot{V}_{STPD}$ until a simulated altitude of approximately 5470 m was reached. Beyond that altitude an increasing hyperventilation was observed. In the acclimatized man, however, the respiratory response at rest and during exercise is different.

Acclimatization to altitude is not only a process of weeks and months but of years or even generations. This became dramatically evident when the

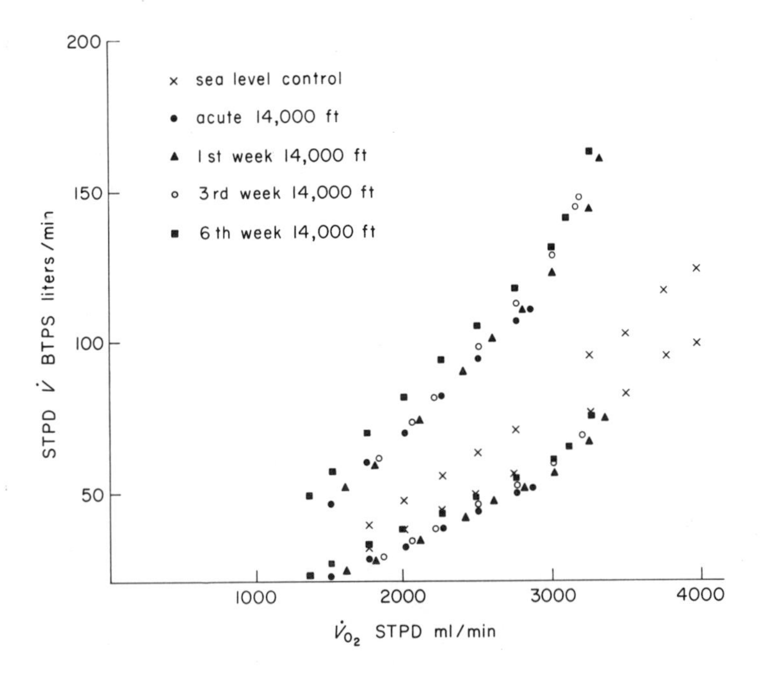

FIG. 2. Relation of oxygen consumption during various levels of exercise to pulmonary ventilation (BTPS and STPD conditions) in acute and subacute/subchronic exposure to the altitude of 14,000 ft (4260 m).

respiratory response of temporarily acclimatized sea level residents was compared with that of natives living permanently at elevations above 13,500 ft. Both groups were studied during exercise over their range of aerobic work tolerance. At any given work intensity, expressed in oxygen demands, the ventilation (BTPS) of the natives was considerably lower than that of the temporarily acclimatized men (Fig. 5). The relative difference in ventilation was approximately 2 : 3. The United States Air Force low pressure chamber installed in Dr. Hurtado's facilities in Morococha, Peru (4420 m) permitted a comparison of the respiratory response to gradually increasing hypoxia between native residents and temporarily acclimatized visitors from sea level. The testing procedure provided a relatively quick ascent of 600 m every 4 min, allowing for a 3 min period of stay at each level. During the second half of such a leveling-off period V_E, $P_{A_{O_2}}$, and $P_{A_{CO_2}}$ were measured by Dr. Velasquez (the San Marcos University, Lima). Up to a simulated altitude of about 6600 m or a $P_{I_{O_2}}$ or 56 mm Hg, there was only a slight difference in the ventilatory response between the native and the temporarily acclimatized sojourner (as Fig. 6 illustrates). However, at the peak altitude of 9730 m, which was attained by both subjects,

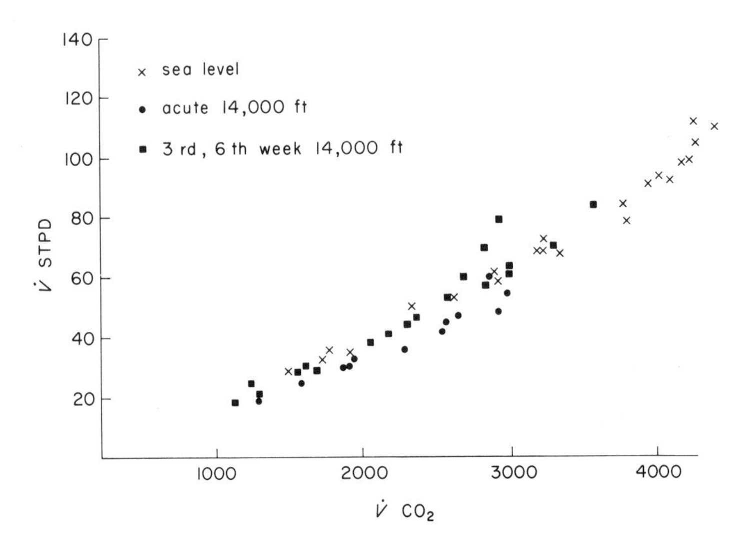

FIG. 3. CO_2 *production during exercise and pulmonary ventilation (STPD) in altitude exposure.*

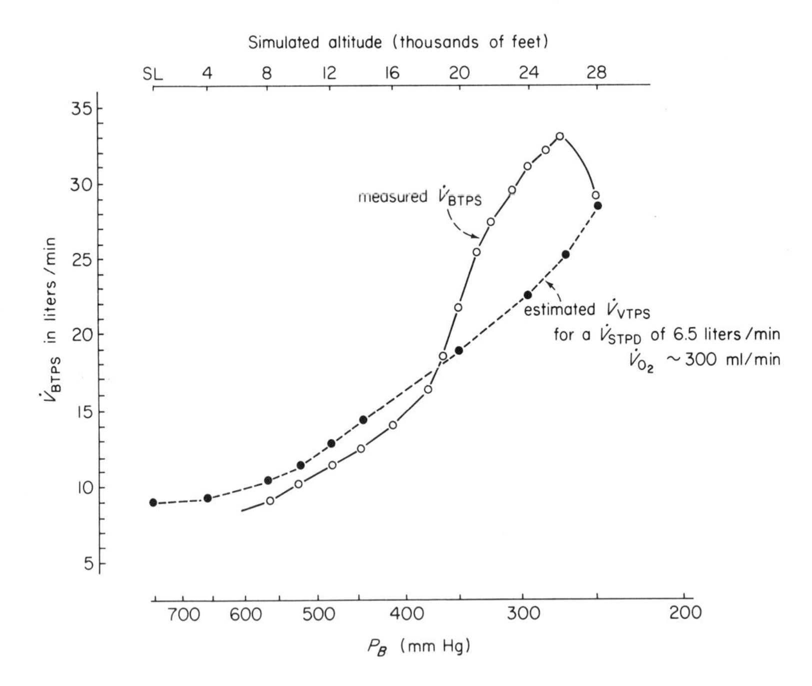

FIG. 4. Theoretical and measured values of pulmonary ventilation (BTPS) for a resting subject, exposed to increasing levels of hypoxia.

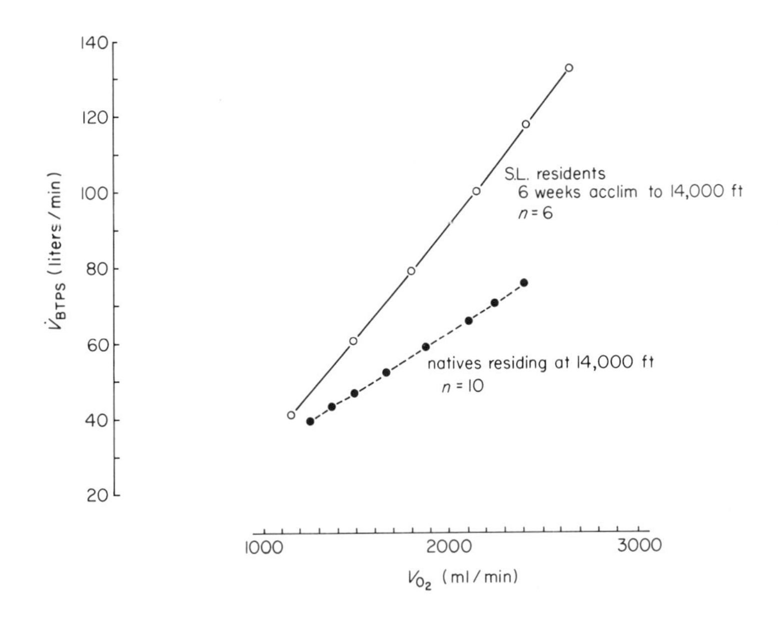

FIG. 5. Comparison of pulmonary ventilation (BTPS) during exercise at altitude of 14,000 ft (4260 m) between temporarily acclimatized sea level residents and altitude natives.

the visitor was breathing at eight times base level ventilation to keep the $P_{A_{O_2}}$ above the critical level of 28 mm Hg, while the native needed only to triple his resting ventilation to stay above his critical $P_{A_{O_2}}$ of 22 mm Hg. The greater ventilatory rate of the sojourner resulted in a fall of $P_{A_{CO_2}}$ to 10 mm Hg.

Previously one had assumed that such hyperventilation could cause severe hypocapnic symptoms and considerable cerebral vasoconstriction, a situation not very favorable to hypoxic tolerance. However, it has been shown (25) that there is adaptation to chronic hyperventilation, leading to a tolerance of very low alveolar carbon dioxide tensions. This was also demonstrated in another chamber experiment at Morococha in which the chamber pressure was lowered in steps as before to a simulated altitude of 9120 m (Fig. 7). At this level a P_B of 225 mm Hg was maintained for 30 min. The subject was the same temporarily acclimatized visitor normally living at sea level. His $\dot{V}_{BTPS}$ during the 30-min period at 9120 m averaged 47 liters/min, which was about five times the base level. With such hyperventilation the $P_{A_{O_2}}$ leveled at 31 mm Hg and the $P_{A_{CO_2}}$ at 11 mm Hg. The subject remained fully conscious and without hypocapnic symptoms. The test was brought to a ceiling with further stepwise lowering of the chamber pressure to a simulated altitude of 10,030 m. The "critical" $P_{A_{O_2}}$ at this point was 24.5 mm Hg, the corresponding $P_{A_{CO_2}}$ was 10 mm Hg. In that

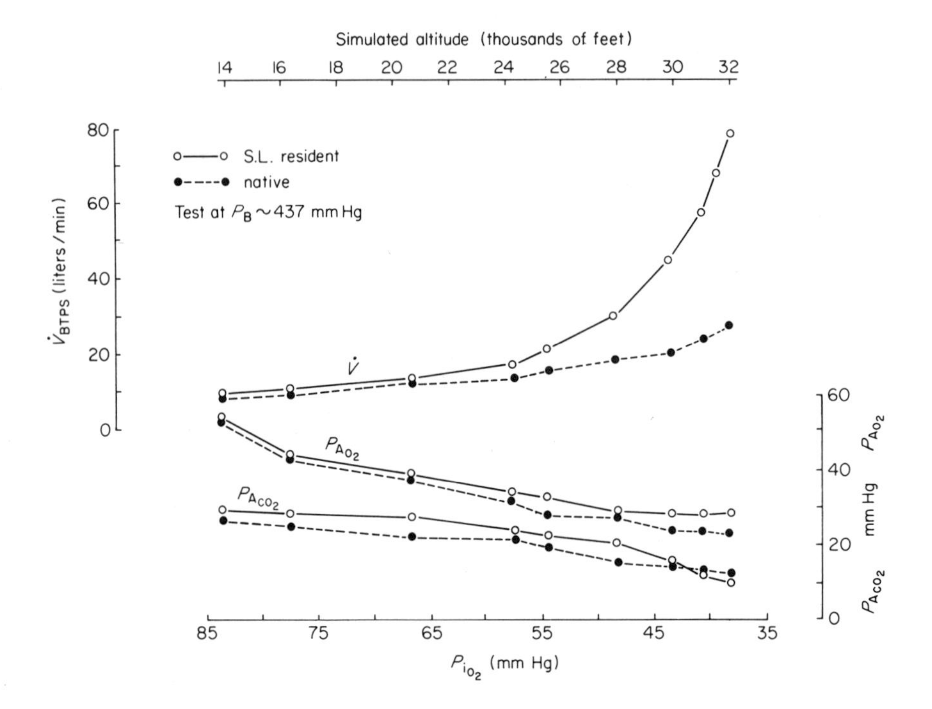

FIG. 6. Comparison of the respiratory response to increasing hypoxia between an altitude native and a temporarily acclimatized sea level resident. Test performed at P_B ~437 mm Hg.

case the adaptation to altitude was not only accomplished by ventilatory adjustments but also with a lowering of the "critical" alveolar O_2 tension (from 30 to 24.5 mm Hg). The lowest $P_{A_{O_2}}$ observed in natives of the "critical" level was ~ 21 mm Hg. This underscores the point Dr. Luft has made previously (Chapter X) that in addition to adaptive changes in blood, respiration and circulation chronic hypoxia also affects oxydative mechanisms at the cellular level.

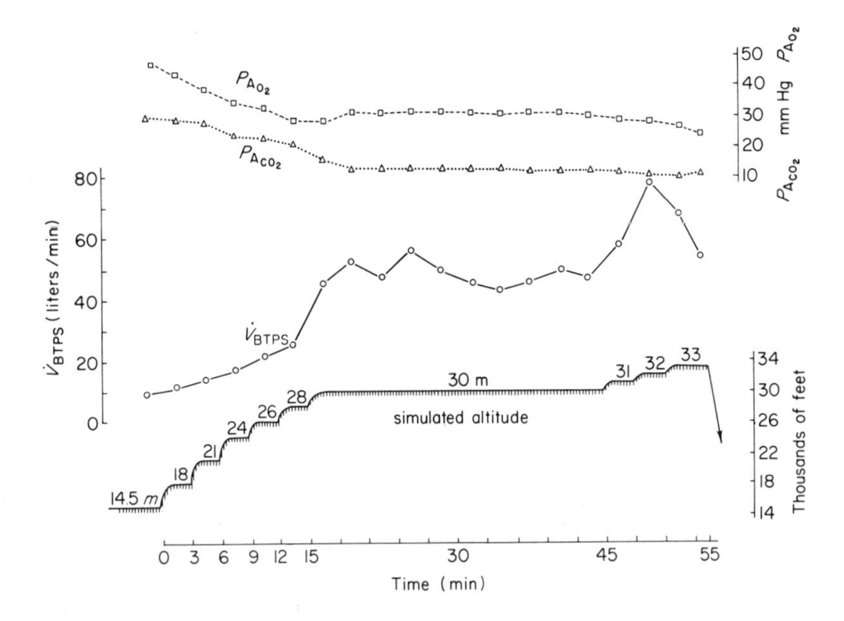

FIG. 7. The respiratory gas exchange of a sea level resident temporarily acclimatized to an altitude of 15,000 ft (4560 m) during an extended exposure (6 weeks) to a simulated altitude of 30,000 ft (9120 m). Altitude tolerance test at P_B = 437 mm Hg.

References

1. Douglas, C. G., Henderson, Y., and Schneider, E. C. (1913). Physiological observations made on Pikes Peak, Colorado, with special reference to adaptation to low barometric pressure. *Phil. Trans. Roy. Soc. London, Ser. B* **203**, 185.
2. Barcroft, J. (1927). "Die Atmungsfunktion des Blutes." Springer-Verlag, Berlin and New York.
3. Herxheimer, H., Kost, R., and Ryjaczek, K. (1933). Untersuchungen ueber den Gasaustausch im Hoehenklima bei leichter und schwerer Muskelarbeit. *Arb. Physiol. Angem. Entomol. Berlin-Dahlem* 7, 308.
4. Balke, B., (1944). Energiebedarf im Hochgebirge. *Klin. Wochenschr.* **23**, 223.
5. Delius, L., Opitz, E., and Schoedel, W. (1942). Hoehenanpassung am Monte Rosa. I. Ruheversuch. *Luftfahrt-Med. Abh.* **6**, 213.
6. Hasselbach, H. A., and Lindhard, J. (1911). Zur experimentellen Physiologie des Hoehenklimas. *Skand. Arch. Physiol.* **25**, 361.
7. Ewig, W., and Hinsberg, K. (1930). Kreislaufstudien im Hochgebirge. *Klin. Wochenschr.* **2**, 1812.
8. Kestner, O., and Schadow, N. (1927). Strahlung, Atmung und Gaswechsel Versuche am Jungfraujoch. *Pfluegers Arch. Gesamte Physiol. Menschen Tiere* **217**, 492.
9. Becker-Freyseng, H., Loeschke, H. Luft, U. and Opitz, E. (1942). Hoehenanpassung am Jungfraujoch: Untersuchungen der Atmung und des Blutes unter Ruhebedingungen. *Luftfahrt-Med. Abh.* **7**, 160.

10. Zuntz, N., Loewy, A., Mueller, F., and Caspari, W. (1906). "Hoehenklima und Bergwanderungen in ihren Wirkungen auf den Menschen." Berlin.
11. Jaquet, A. (1925). Stoffwechselvorgaenge bei herabgesetztem Luftdruck. *Schweiz. Med. Wochenschr.* **1**, 755.
12. Grover, R. J. (1963). Basal oxygen uptake of man at high altitude. *J. Appl. Physiol.* **18**, 909.
13. Dill, D. B. (1966). Hypoxia: High altitude revisited. *U.S. Pub. Health Serv. Publ.* **999-AP-25**.
14. Grover, R. F., Reeves, J. T., Will, D. H., and Blount, S. G., Jr. (1963). Pulmonary vasoconstriction in steers at high altitude. *J. Appl. Physiol.* **18**. 567.
15. Reeves, J. T., Grover, E. B., and Grover, F. R. (1963). Pulmonary circulation and oxygen transport in lambs at high altitude. *J. Appl. Physiol.* **18**, 560.
16. Ewig, W., and Hinsberg, K. (1931). Kreislaufstudien. III. Beobachtungen im Hochgebirge. *Z. Klin. Med.* **115**, 732.
17. Dill, D. B., Foelling, A., Oberg, S. A., Pappenheimer, A. M., Jr., and Talbott, J.H. (1931). Adaptations of the organism to changes in oxygen pressure. *J. Physiol. (London)* **71**, 47.
18. Christensen, E. H. (1937). Sauerstoffaufnahme und respiratorische Funktionen in grossen Hoehen. *Skand. Arch. Physiol.* **76**, 88.
19. Schilling, J. A., Harvey, R. B., Becker, E. L., Velasquez, T., Wells, G., and Balke, B. (1956). Work performance at altitude after adaptation in man and dog. *J. Appl. Physiol.* **8**, 381.
20. Boothby, W. M., Lovelace, W. R., II, Benson, O. O., Jr., and Strehler, A. F., (1954). Volume and partial pressures of respiratory gases at altitude. *In* "Handbook of Respiratory Physiology" (W. Boothby, ed.), p. 39. Air Univ., Sch. Aviat. Med., USAF, Randolph Air Force Base, Texas.
21. Rahn, H., and Otis, A. B. (1949). Men's respiratory response during and after acclimatization to high altitude. *Amer. J. Physiol.* **157**, 445.
22. Riley, R. L., Otis, A. B., and Houston, C. S. (1954). Respiratory features of acclimatization to altitude. *In* "Handbook of Respiratory Physiology" (W. Boothby, ed.), p. 143. Air Univ., Sch. Aviat. Med., USAF.
23. Purves, M. S. (1969). Changes in oxygen consumption of the carotid body of the cat. *J. Physiol. (London)* **200**, 132.
24. Dubois, R. M., and Yamamoto, W. (1969). Modification of ventilatory response of crossed perfused rat by stirring the blood stream. *Fed. Proc., Fed. Amer. Soc. Exp. Biol.* **28**, 338.
25. Balke, B., Ellis, J. P., and Wells, J. G. (1958). Adaptive response to hyperventilation. *J. Appl. Physiol.* **12**, 269.

XV

Vertebrates at Altitudes

R. W. BULLARD

Introduction

This summer will mark the 36th anniversary of the publication entitled "Comparative Physiology in High Altitudes" by F. G. Hall, D. B. Dill, and E. S. Guzman Barrón (1). Their paper represented the first real studies performed in the context of modern physiology on birds and mammals native to high-altitude environments. The presented data, as I shall point out, are now resurrected to provide a needed stimulus for rethinking our concepts of the role of hemoglobin in hypoxia.

Comparative physiology of vertebrates in high altitudes is an area that has not been thoroughly investigated. In 1964 the late Dr. Raymond Hock reviewed the existing literature on amphibians and reptiles at high altitude (2). Because very little information has been added since this review, and because physiological information on those fish inhabiting high mountain lakes is not very abundant the present report is confined primarily to mammals, and secondarily to birds, native to those altitudes sufficient to impose hypoxic problems.

Existence at high altitude may involve strong ultraviolet radiation, extremes of temperature, high winds, dehydration, and lack of food supply, as well as hypoxia. According to Morrison wild mammals and birds have successfully met the handicaps of hypoxia to live in any environment in which food is available (3). Quite often barren mountain regions are surprisingly rich in fauna in both the variety of species and the number of individuals (4).

Many species in the Andes inhabit regions to 5200 m (4) and in the Himalayas to 6000 m (5), the upper limit perhaps being related more to the lack of food than to hypoxia. Many animals are capable of making forays to higher altitudes for food, escape, or migration. The champion of all must be the bar-headed goose which reportedly migrates over the high Himalayas from India to the lakes of Tibet at an altitude of 10,000 m (6). Here the ambient P_{O_2} approximates 50 mm Hg, and one is forced to wonder how the hypoxic goose can make any progress. Numerous reports have been made by mountaineering expeditions of avian activity at least to 8200 m (5), and some physiological studies have been made of flight at a simulated altitude of 6100 m (7). Eggs of the snow partridge have been found at 5800 m representing what appears to be an upper limit for a vertebrate life cycle (5).

Most physiological investigation has been done on species living between 3300 and 4550 m. Further investigation utilizing critical experiments is required, but at least we are at a point where the important questions can be asked and some even answered, at least in part. Has altitude life for many generations forced a different set of physiological responses than those of the brief sojourner? Has the adaptation of the native animal been in the direction of maintaining a higher tissue P_{O_2}, or is the adaptation such that the organism maintains function in spite of a lower tissue P_{O_2}.

After carefully considering the comparative data it appears that the general dogma we physiologists preach about adaptation to the chronic hypoxia of high altitude has been ignored by those organisms proven most fit for life in such environments.

Function in Oxygen Lack and the Critical P_{O_2}

A simple question that serves as our starting point is whether or not the high-altitude native mammal can maintain function at a lower P_{O_2} than can sea-level animals either adapted to high altitude or not. Figure 1 represents a series of experiments performed at the Barcroft Laboratory of the White Mountain Research Station at 3800 m on the laboratory rat (Long-Evans), born at Barcroft and the golden mantled ground squirrel (*Citellus lateralis*) of this altitude and of similar body size (8). The functions of temperature regulation, oxygen consumption, heart rate, and cardiac output were followed as the inspired P_{O_2} was shifted from 101 mm Hg, the ambient level at Barcroft, to 49 mm Hg, approaching a simulated altitude of 10,000 m. With this decrement in P_{O_2} the ground squirrel was better able to maintain colonic temperature and oxygen consumption. The difference in the rate of cooling was not due to fur

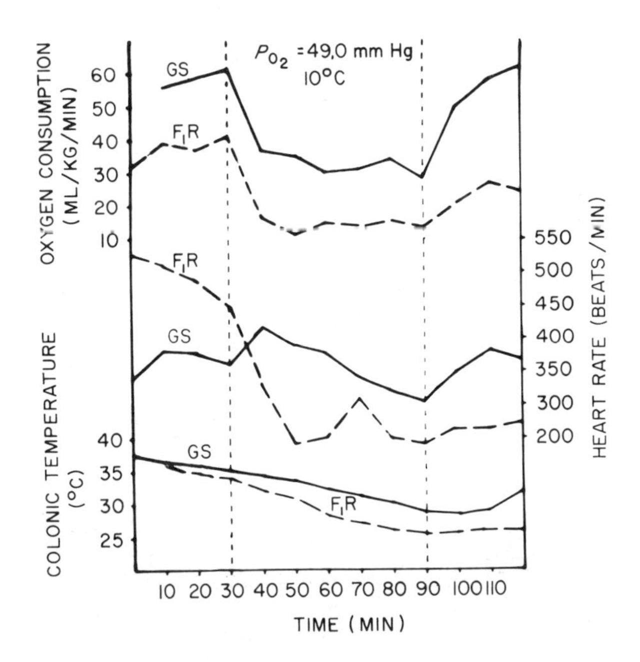

FIG. 1a. Alteration in body temperature, heart rate, and oxygen consumption in Citellus lateralis *of 3800 m and the laboratory rat born at 3800 m as inspired P_{O_2} is changed from 100 to 49 mm Hg at 10°C (8).*

insulation differences since the rat in these experiments had slightly better insulation. The higher body temperature was then the result of higher oxygen consumption or greater ability to maintain heat production. Because the oxygen consumptions are higher than what would be predicted for an animal this size, even with the P_{O_2} of 49 mm Hg, thermogenic activity was continued (7).

The ground squirrel responded to the hypoxic test by increasing its heart rate and cardiac output, whereas those functions in the rat fell drastically. Most of the high altitude adapted rats died before the 60 min of exposure to a P_{O_2} of 49 mm Hg at 20°C was completed.

These results clearly indicate a difference in function at lowered ambient P_{O_2} levels between the high-altitude native ground squirrel and the rat born at altitude and presumably acclimatized. Morrison was able to precisely quantify the P_{O_2} at which oxygen transport or oxygen uptake during moderate cold exposure was reduced, or the so-called "critical P_{O_2}," in a series of lowland and altiplano rodents (3) native to approximately 4200 m. These results indicated little overlap between the "critical P_{O_2}" of the lowland species (mean = 93 mm Hg) and their altitude relatives (mean = 68 mm Hg). In one series of experiments

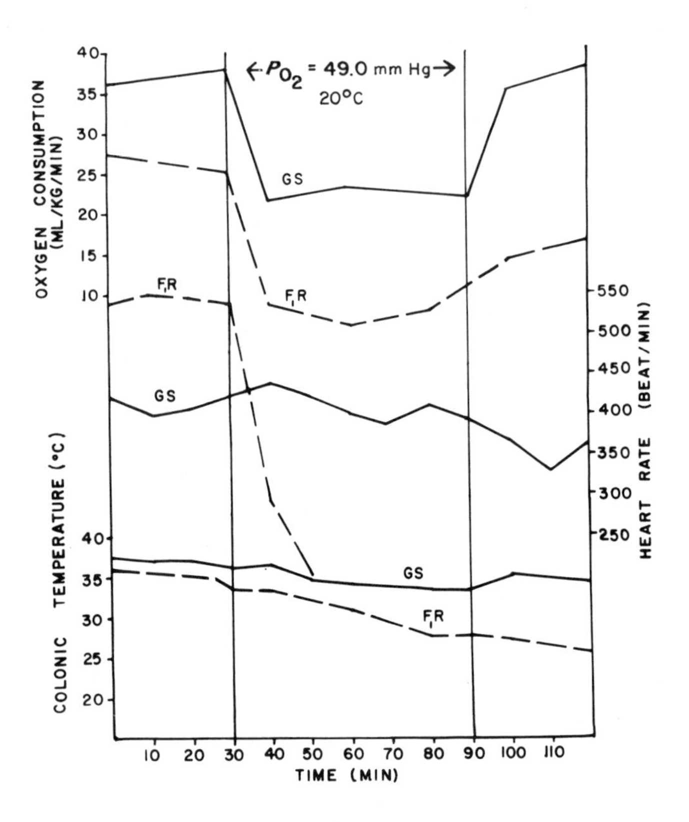

FIG. 1b. Same test as in Fig. 1a but at 20°C (8).

more similar to those performed by our group, Morrison showed that the wild guinea pig of the altiplano possessed a much lower critical P_{O_2} than did the domesticated variety of guinea pig transposed to the same altitude (3). It appears that hypoxic tolerance is thus enhanced by a species' history in high-altitude environments.

Regulatory Mechanisms or Systems Adaptation

The greater functional maintenance in hypoxia for the ground squirrel or the lower "critical P_{O_2}" of the highland rodent must either be due to maintenance of a higher tissue P_{O_2} or to tissue adaptation wherein function is continued with a P_{O_2} reduction. The maintenance of functional tissue P_{O_2} levels in the face of hypoxia involves respiratory and cardiovascular functions, the oxygen carrying

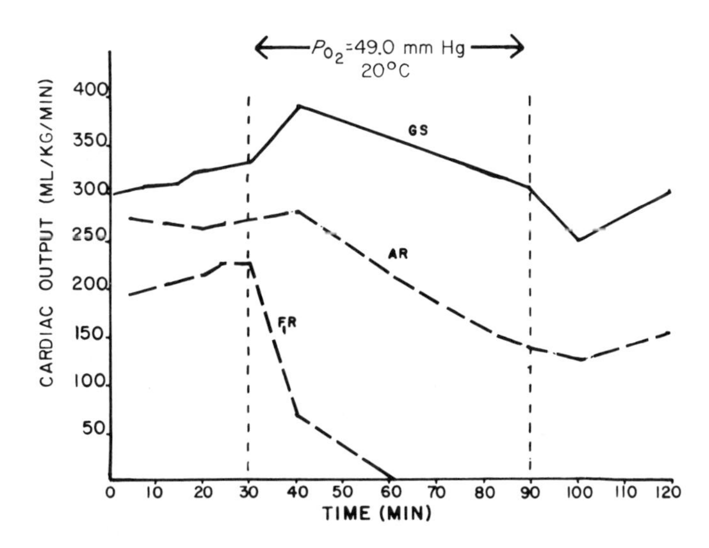

FIG. 1c. Alteration in cardiac output (by direct Fick method) as inspired P_{O_2} was changed from 100 to 49 mm Hg at 20°C (8).

capacity of the blood, and characteristics of the oxygen hemoglobin equilibrium curve. The problem of intracellular P_{O_2} determination has not been solved, and adequate and direct measurements for high altitude animals thus do not exist. Figure 2 shows for the rat and ground squirrel at Barcroft and at the equivalent of 10,000 m of altitude the P_{O_2} tensions from inspired air to venous blood. These values were computed from blood oxygen contents obtained in experiments as shown in Fig. 1 and the oxygen hemoglobin equilibrium curve (9) for both species. Because little information was available on the Bohr factors or blood pH the values represent only approximations.

The arterial P_{O_2} was higher in the ground squirrel than in the rat both at Barcroft and simulated 10,000 m. This difference was most likely due to more effective ventilatory responses in the ground squirrel assuming that the alveolar-arterial gradient was not increased at altitude. It has been demonstrated that in contrast to the human highland native (10), other native species have substantial hypoxic ventilatory drives (11). However, Morrison and Elsner in studying the respiratory rate alone in hypoxia concluded that most high-altitude rodents did not show rate increases to hypoxia (12). It is obvious that this is an experimental approach requiring further development.

At 3800 m the calculated mean capillary P_{O_2}'s were quite similar in both the rat and ground squirrel. However, the conclusion that cellular P_{O_2} levels were the same cannot be made without more substantial knowledge of comparative intercapillary distances. At the simulated altitude of 10,000 m the capillary P_{O_2}

of the rat was considerably lower. These animals were suffering from severe functional deterioration with failure of the respiratory and circulatory systems to deliver adequate oxygen. When arterial P_{O_2} values less than 20 mm Hg were reached in the rat, death of the animal resulted (8). This value was not reached in the ground squirrel and consequently all animals survived the hypoxic test.

At this point it would appear that in the more severe hypoxia the ground squirrel was better able to maintain systems function and thereby maintain oxygen delivery. The mechanism that permitted this, however, was probably a tissue-level adaptation, whereby the important functions, particularly of the myocardium and the respiratory muscles, was continued. This is a point that has also been made by Burlington for this same species of ground squirrel (13).

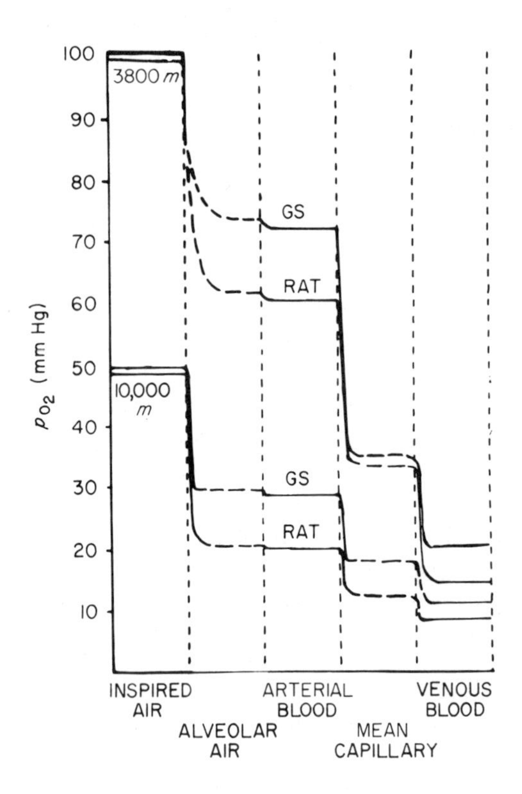

FIG. 2. The oxygen pressure gradient from inspired air to venous blood in Citellus lateralis *at 3800 m and the laboratory rat born at 3800 m for ambient oxygen at P_{O_2} of 100 mm Hg (3800 m) and 49 mm Hg (10,000 m). [Values approximated from published data (8, 9).]*

BLOOD CHARACTERISTICS AND HEMOGLOBIN

The parameters of high altitude adaptation that have received the most attention are the augmented hematopoietic activity and the resulting elevated red blood cell and hemoglobin concentrations. This is an alteration of adaptive advantage as a greater amount of oxygen can be carried at partial saturations and adequate unloading can occur without requiring marked depression of P_{O_2} levels in the tissue. Interestingly enough a survey of the literature as has been done for Table I, on the native or wild highland animals, clearly indicates that this is a form of adaptation that has not been used.

The adapted rabbit and rat, two sea-level forms (1), are included in the table for the purpose of comparison, and at altitude these forms have the highest values of hemoglobin concentrations. Llamas at sea level had a higher oxygen capacity than did llamas at 5340 m. All of the altitude species shown have quite normal blood pictures, or, in some cases, slightly on the anemic side. Morrison concluded on the basis of many more species than are shown here that the altitude of origin and hematocrit ratio do not correlate (14, 15).

As Table I indicates, red blood cells vary greatly in size and hemoglobin content. Because of increased surface area for oxygen diffusion a smaller cell as found in the Camelidae may have adaptive benefit in presenting increased surface area for oxygen diffusion. This appears to be a characteristic of the family rather than a specific altitude adaptation (16).

Three other species not shown in Table I deserve special comment. These are (a) Apodemus, the Russian wood mouse (17); (b) Peromyscus, the deer mouse of White Mountain (18); and (c) the human (19). All three forms show increased hemoglobin concentration at high altitude, and all three forms could be considered high-altitude natives. However, as Morrison (3) has emphasized these forms are not members of isolated highland populations, and there is probably a continuous genetic admixture from the surrounding lowlands.

With these three exceptions hemoglobin and hematocrit ratio increases do not appear to yield any selective advantages. Increased hematocrit ratios may impose a serious hemodynamic penalty (20, 21). Barbarshova (22) has contended that the polycythemia of altitude exposure is a pathological sign rather than a beneficial adaptation.

OXYGEN HEMOGLOBIN EQUILIBRIUM CURVE

The concepts and evaluations of the position of the equilibrium curve in adaptation to a hypoxic environment have gone through several trends of thought. The latest arises from the results of careful biochemical investigations of red blood cell metabolism. There are several facts that need be noted.

TABLE I. Blood Characteristics of Altitude Dwellers[a]

Species	*Altitude or range*	*Ref.*	*Hb conc (gm/100 ml)*	*Hematocrit (ratio %)*	*RBC count (millions/mm^3)*	*Mean cell[b] Hb(mmg)*	*RBC[b] size (μ^3)*
Rat	0	(9)	15.0	45	7.91	19.3	56
Rat	3800	(9)	19.3	59	9.81	20.2	55
C. lateralis	3800	(9)	14.3	48	6.93	20.7	70
M. flaviventris	3800	(9)	15.5	49	7.72	20.2	64
Chinchillula	3900	(15)	13.3	42	7.00	19.0	59
Hesperomys (laucha)	4540	(15)	11.9	43	9.00	13.2	48
Phyllotis posticalis (pericot)	4540	(15)		35			
Akodon[c] (vole mouse)	3900-4540	(15)	15.3	48	10.05	15.2	47
Vicuna	2810	(1)	12.6	30	14.05	8.6	21
Vicuna	4710	(1)	13.6	32	16.61	8.2	19
Viscacha	3660	(1)	11.1	32	7.12	15.6	45
Llama	0	(1)	17.5	39	11.43	15.3	45
Llama	5340	(1)	11.2	26	11.10	10.1	23
Rabbit	5340	(1)	16.7	57	7.00	23.9	82
Rabbit	0	(1)	11.7	35	4.55	25.8	78

[a]See text for comment on exceptions. [b]Calculated from measured values. [c]Average of 4 species and 15 values.

1. In the metabolism of glucose to pyruvate in the red blood cell the pathway may proceed from 1,3 diphosphoglycerate (1,3-DPG) to 3-phosphoglycerate (3-PG) or follow a secondary route of 1,3-DPG to 2,3-DPG to 3-PG (23, 24).

2. In conditions of hypoxia and possibly increased pH as may occur temporarily at altitude the concentration of 2,3-DPG in the red blood cell increases. This has been measured at about 10-20% at Leadville and 20% at Morococha (25, 26). Increases have also been shown for the rat and domesticated guinea pig at high altitude (27). In some hypoxic diseases the increases are greater (23).

3. Reduced hemoglobin has a tendency to bind 2,3-DPG; therefore with the hypoxia of altitude the binding would be enhanced. The effect of the binding is to reduce the affinity of hemoglobin for oxygen or to produce a rightward shift in the curve (23). The details of the molecular chemistry of the binding has been recently described by Perutz (28).

Within slightly more than a 1-year period at least 10 articles have appeared on the "essential" role of 2,3-DPG in hypoxic adaptations. Without trying to dampen this enthusiasm for a new molecule, it should be pointed out that there may be some powerful lessons from comparative physiology. Figure 3 is a conglomerate of data from Schmidt-Nielsen and Larimer (29) and Moll and Bartels (20) showing the relationship of the logarithm of the P_{50} or oxygen

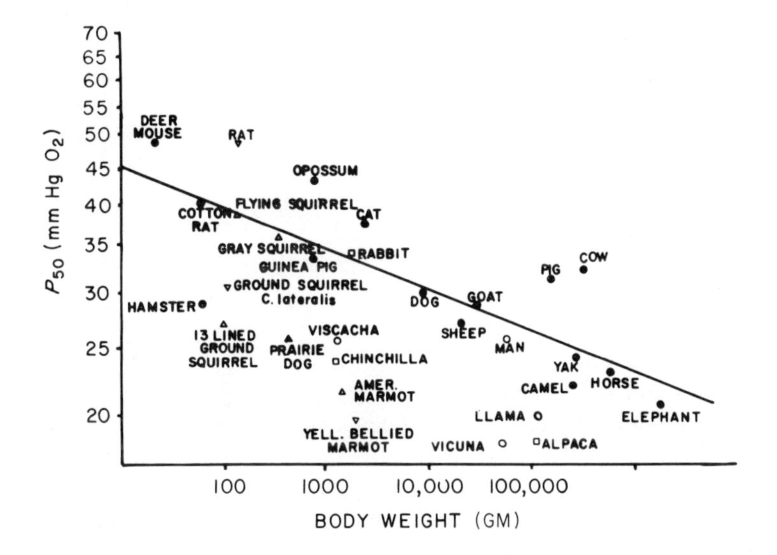

FIG. 3. The logarithm of P_{50} (mm Hg) or oxygen tension at which 50% saturation of hemoglobin is obtained versus logarithm of body weight. The primary data are from Schmidt-Nielsen and Larrimer (29) and Moll and Bartels (20). Points indicated by ∇ are from Bullard, Broumand, and Meyer (9); by □, from Chiodi (34), by △, from Hall (31), and by ○, Hall, Dill, and Barrón (1).

tension required for half saturation, to the logarithm of body weight. To this plot we have added all of the P_{50} values that we could find for the native high-altitude species. All of these values fall below the Schmidt-Nielsen and Larimer line, indicating a shift to the left of the equilibrium curve for mountain natives. Admittedly these values come from a variety of laboratories and were determined with differing techniques, yet lack of exception is impressive. Those data (9) obtained in our own studies on *Citellus lateralis* and *Marmota flaviventris* were made on fresh blood and contrast markedly to that of the rat also shown. This we measured at high altitude on blood probably containing increased 2,3-DPG. It thus appears from Fig. 3 that those animals that have most successfully adapted to an hypoxic existence are probably not depending on 2,3-DPG.

In point of fact some of the most successful residents of altitude possess blood which does not contain 2,3-DPG (30). These are the hoofed members of the order Artiodactyla which include the yak and Camelidae, the alpaca, the vicuna and the llama. The sheep has had a history of altitude success and possesses a hemoglobin that does not bind 2,3-DPG (28).

In Fig. 3 some of the data of Hall (31) on members of the squirrel family not altitude residents are included. The burrowing animals, the 13 lined ground squirrel, the prairie dog, and the American marmot are presumably adapted to hypoxia with curves to the left, and can be contrasted to the related gray squirrel and flying squirrel that do not face hypoxic challenges.

Of much interest here is the demonstration that the leftward curves are not unique to altitude species. It is possible that the burrowing rodent and the Camelidae possess special adaptive characteristics that enable these forms to move successfully into the hypoxic environment of high altitude. The evolution of the leftward curve does not pose a serious challenge as most mammals possess such a hemoglobin during fetal and early neonatal existence (32). Fetal hemoglobin represents an adaptation to hypoxia, yet it is bound only weakly by DPG.

Those mammals that have a curve to the left tolerate hypoxia better (33, 34). This negative correlation of P_{50} and hypoxic tolerance has been forgotten with the modern emphasis on DPG.

Both birds studied by Hall, Dill, and Barrón, the Andean goose, (*Chloephaga melanoptera*) and the American ostrich (*Rhea americana*) at an altitude of 3660 m had equilibrium curves to the left of a group of sea-level birds (1). With the many species of birds able to exercise vigorously at altitudes more data must be obtained.

CAPILLARITY AND THE VASCULAR BED

A very rational concomitant adaptation with the leftward oxygen hemoglobin equilibrium curve would be an increased capillary density or a decreased intercapillary distance (35). Reduction of the diffusion pathway length could be the most effective compensation to ensure adequate oxygenation. The most complete study on a native mammal has been done by Valdiva on the native guinea pig of Peru at approximately 4200 m using India ink perfusion (36). The altitude animals had approximately 42% greater number of capillaries per muscle fiber for four different muscles when compared to sea-level guinea pigs. Such changes also occur in adapted sea-level animals, but more physiological data are required on a wider variety of species.

The total vascular bed of the native high-altitude rodent may be larger than its sea-level relative adapted to altitude. Table II shows that the ground squirrel and marmot of high altitude have a greater blood volume than the adapted rat. As can be noted this difference results from a larger plasma volume. In the rodents the red blood cell volume was the same for all three species (9).

Although the total blood volume of the three species of Camelidae are roughly the same as the human or lower when expressed in kilograms of body mass, the plasma volume are much larger (37). These species have a much lower red cell mass, but are by no means functionally impaired.

One may speculate that the rodents probably developed polycythemia upon first inhabiting the mountain areas. This was then compensated for by an increased plasma volume. It is an interesting note that the high-altitude native possesses a greater plasma volume than red blood cell mass, but it is this latter parameter that has received far greater emphasis in the literature.

Adaptation at the Tissue Level

From Figs. 1 and 2 it is suggested that function is continued in the ground squirrel with a markedly reduced arterial capillary and venous P_{O_2}. This would suggest that the tissue may possess the ability to function at low P_{O_2}. Such adaptations could consist of changes in enzymatic activity to more effectively utilize oxygen at a low P_{O_2}, a greater number of mitochondria, and an increased capacity for anaerobic glycolysis (38, 39). This latter aspect has been of interest to us because it can be readily demonstrated.

Figure 4 shows results obtained on isolated tissues by Burlington and Maher (40) on the isolated atrium and by our own laboratory on the isolated phrenic nerve-diaphragm preparation (41). In these experiments, the tissues are stimulated to contract in an aerated physiological solution. At 0 time (ordinate) the aerating oxygen is replaced by nitrogen, and the course of contraction followed. As can be seen altitude acclimation improved the anoxic performance of both the atrial and phrenic nerve-diaphragm preparations from the rat. However, the isolated atrium from *C. lateralis* and the phrenic nerve-diaphragm

TABLE II. Plasma and Blood Volumes of Altitude Dwellers[a]

Species	*Altitude (m)*	*Total blood vol (ml/kg body wt)*	*Plasma vol (ml/kg body wt)*	*Erythrocyte vol (ml/kg body wt)*
Alpaca[b]	4200	72	45	24.3
Llama[b]	4200	65	40	25
Vicuna[b]	4300	87	56	31.3
Human[a,b]	4540	88	33	55
Golden Mantled ground squirrel[c]	3800	113	59	48
Yellow Bellied Marmot[c]	3800	100	51	54
Laboratory rat[b,c]	3800	83	35	50

[a]Sea-level species adapted to high altitude. [b]Data from Reynafarje (37). [c]Data from Bullard *et al.* (9).

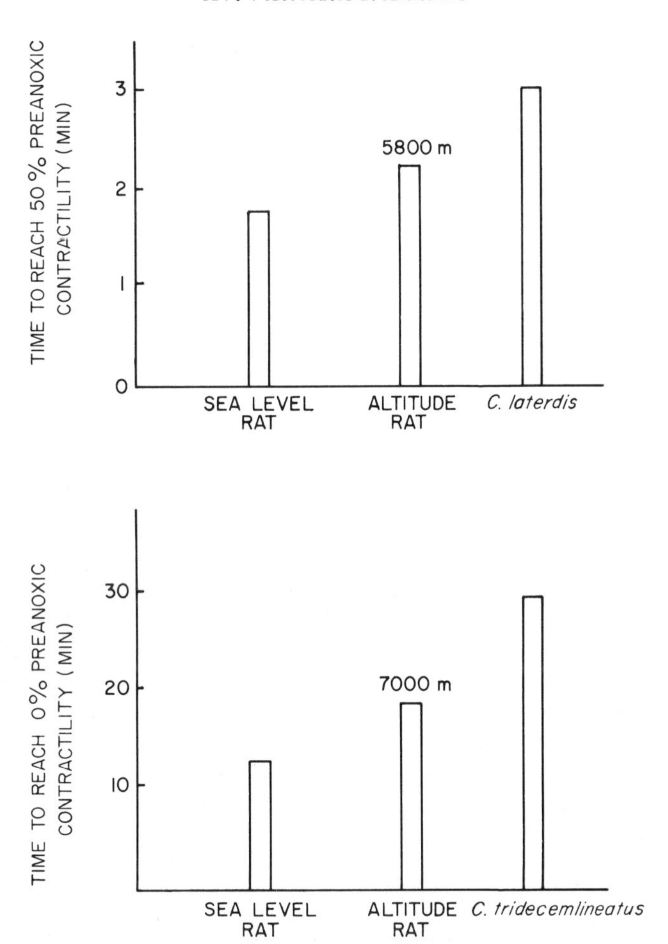

FIG. 4. Evidence for tissue adaptation in the isolated atrium and phrenic nerve-diaphragm preparation. At 0 time on the ordinate the aerating gas in the bathing medium was switched from 95% O_2 and 5% CO_2 to 95% H_2 and 5% CO_2. Atrial data (upper) are from Burlington and Maher (40) with rat acclimated to 5800 m. Atrium was stimulated at 1.5/sec. Phrenic nerve-diaphragm (lower) preparation data from Komives and Bullard (41). The phrenic nerve was stimulated at 20/min.

preparation from *C. tridecemlineatus* continued to function far longer than those from the rat after acclimation. With the two squirrels altitude adaptation had not been imposed, and thus this functional ability is perhaps a characteristic of the hibernating or burrowing mammals. Again, this may be a characteristic that had permitted these animals to move successfully into high altitude.

The increased tolerance may be due to several factors. An increase in the muscle isozyme or M form of lactic dehydrogenase over the H or heart isozyme has been demonstrated for the rat heart after simulated altitude exposure (42). The M form promotes conversion of pyruvate to lactate and thereby facilitates glycolysis. The H isozyme is the aerobic form and catalyzes the reaction of lactate to pyruvate.

A second factor that may be important for the burrowing animal or its relative that has moved to high altitude would be an ability for the tissues to function during pH changes. Recent studies in our laboratory have shown that the isolated hamster heart is much more resistant to artificially induced respiratory acidosis than is the rat heart. This property could be important in preventing both feedback inhibition of glycolysis and cellular damage (43).

Discussion

The physiological data presented herein suggest that the mammalian natives of high altitude studied to date possess certain characteristics that can be very much different from those found in the commonly used laboratory animals or man when adapted to chronic hypoxia. In summary these characteristics are (1) the leftward position of the oxygen hemoglobin equilibrium curve when compared on the basis of the P_{50} body weight relationship curve; (2) absence of increased hematocrit ratio or hemoglobin concentration in the blood; (3) an expanded plasma volume when compared to the sea-level animals in exposure to chronic hypoxia; (4) on the basis of limited evidence, a greater ability for the tissues to function with a low P_{O_2} or under anaerobic conditions; (5) again, on limited evidence, a greater ability to regulate tissue P_{O_2} by maintaining circulatory and respiratory function.

It is perhaps unfortunate that all of the mountain mammals studied extensively are either rodents or Camelidae. The rodent can carry to altitude those adaptations for burrowing and possibly hibernating that may provide fitness in the hypoxic environment. The Camelidae even at low altitude possess rather unusual adaptations as a family (30) such as the leftward oxygen-hemoglobin equilibrium curve. Thus it does not appear that those characteristics of the high altitude mammal are really that unique. Unfortunately there does not exist enough information on resident highland birds.

The lack of high hematocrit ratio in the native animals, after some teleological considerations, is a very reasonable finding. When one analyzes the overall physiology of the animal there must be some reasonable compromise between hematocrit ratio and other hemodynamic parameters. Because these other parameters may not be greatly different between altitude and sea level, one would expect the adapted resident to also establish an optimal hematocrit ratio.

At this point it is entirely compatible with existing data to divide mammals into two groups. First would be the typical normoxic type that upon encountering hypoxia responds by an increase in the hemoglobin and hematocrit ratio and possibly by a rightward shift in the oxygen-hemoglobin equilibrium curve through the 2,3-diphosphoglycerate effect. The second group would be the type that are the successful altitude dwellers, the rodents and Camelidae and other less studied artiodactyl. These animals do not increase the hemoglobin concentration or shift their oxygen equilibrium curve to the right. Most mammals exist in this form during fetal and neonatal life.

It would be tempting but far too simple to conclude that the mammal of high altitude is one that never grew up. However, all of the adaptations present in the native high altitude animal appear to be present in sea-level animals at least in one stage of the life cycle.

References

1. Hall, F. G., Dill, D. B., and Guzman Barrón, E. S. (1936). Comparative physiology in high altitudes. *J. Cell Comp. Physiol.* **8**, 301.
2. Hock, R. J. (1964). Animals in high altitudes: Reptiles and amphibians. *In* "Handbook of Physiology" (Amer. Physiol. Soc., J. Field, ed.), Sect. 4, Chapter 33, p. 841. Williams & Wilkins, Baltimore, Maryland.
3. Morrison, P. (1964). Wild animals at high altitudes. *Symp. Zool. Soc. London* **13**, 49.
4. Peason, O. P. (1951). Mammals in the highlands of Southern Peru. *Bull. Mus. Comp. Zool., Harvard Univ.* **106**, 117.
5. Swan, L. W. (1961). The ecology of the high Himalayas. *Sci. Amer.* **205**, 68.
6. Swan, L. W. (1970). Goose of the Himalayas. *Natur. Hist., N.Y.* **79**, 68.
7. Tucker, V. A. (1967). Respiratory physiology of house sparrows in relation to high-altitude flight. *J. Exp. Biol.* **48**, 55.
8. Bullard, R. W., and Kollias, J. (1966). Functional characteristics of two high-altitude mammals. *Fed. Proc., Fed. Amer. Soc. Exp. Biol.* **25**, 1288.
9. Bullard, R. W., Broumand, C., and Meyer, F. R. (1966). Blood characteristics and volume in two rodents native to high altitude. *J. Appl. Physiol.* **21**, 994.
10. Severinghaus, J. W., Bainton, C. R., and Carcelen, A. (1966). Respiratory insensitivity to hypoxia in chronically hypoxic man. *Resp. Physiol.* **1**, 308.
11. Brooks, J. G., and Tenney, S. M. (1968). Ventilatory response of llama to hypoxia at sea level and high altitude. *Resp. Physiol.* **5**, 269.
12. Morrison, P., and Elsner, R. (1962). Influence of altitude on heart and breathing rates in some Peruvian rodents. *J. Appl. Physiol.* **17**, 467.
13. Burlington, R. F., Maher, J. T., and Sidel, C. M. (1969). Effect of hypoxia on blood gases and acid base balance in *in vitro* myocardial function in a hibernator and nonhibernator. *Fed. Proc., Fed. Amer. Soc. Exp. Biol.* **28**, 1042.
14. Morrison, P. R., Kerst, K., and Rosenmann, M. (1963). Hematocrit and hemoglobin levels in some Chilean rodents from high and low altitude. *Int. J. Biometeorol.* **7**, 45.
15. Morrison, P. R., Kerst, K., Reynafarje, C., and Ramos, J. (1963). Hematocrit and hemoglobin levels in some Peruvian rodents from high and low altitude. *Int. J. Biometeorol.* **7**, 51.
16. Bartels, H., Hilpert, P., Barbey, K., Betke, K., Riegel, K., Lang, E. M., and Metcalf, J.

(1963). Respiratory functions of blood of the yak, llama, camel, Dybowski deer, and African elephant. *Amer. J. Physiol.* **205**, 331.

17. Kalabukov, N. J. (1937). Some physiological adaptations of the mountain and plain forms of the wood mouse (*Apodemus sylvaticus*) and of other species of mouse like rodents. *J. Animal Ecol.* **6**, 254.
18. Hock, R. J. (1964). Physiological responses of deer mice to various native altitudes. *In* "The Physiological Effects of High Altitude" (W. H. Wiehe, ed.), p. 59. Pergamon, Oxford.
19. Hurtado, A. (1964). Animals in high altitude: Resident man. *In* "Handbook of Physiology" (Amer. Physiol. Soc., J. Field, ed.), Sect. 4, p. 843. Williams & Wilkins, Baltimore, Maryland.
20. Moll, W., and Bartels, H. (1968). Oxygen binding in the blood of mammals. *In* "Oxygen Transport in Blood and Tissue" (D. W. Libben *et al.*, eds.), p. 39. Thieme, Stuttgart.
21. Smith, E. E., and Crowell, J. W. (1963). Influence of hematocrit ratio on survival of unacclimatized dogs at simulated altitude. *Amer. J. Physiol.* **205**, 1172.
22. Barbashova, Z. I. (1958). Specific and non specific lines of acclimatization to hypoxia. *In* "Problems of Evolution of Physiological Function," p. 116. Acad. Sci. U.S.S.R. Moscow.
23. Brewer, G. J., ed. (1970). "Red Cell Metabolism and Function." Plenum Press, New York.
24. Brewer, G. J., and Eaton, J. W. (1971). Erythrocyte metabolism: Interaction with oxygen transport. *Science* **171**, 1205.
25. Eaton, J. W., Brewer, G. J., and Grover, R. F. (1969). Role of red-cell 2,3-diphosphoglycerate in the adaptation of man to altitude. *J. Lab. Clin. Med.* **73**, 603.
26. Torrance, J. D., Lenfant, C., Cruz, J., and Marticorena, E. (1970-1971). Oxygen transport mechanisms in residents at high altitude. *Resp. Physiol.* **11**, 1.
27. Baumann, R., Bauer, C., and Bartels, H. (1971). Influence of chronic and acute hypoxia on oxygen affinity and red cell 2,3 diphosphoglycerate of rats and guinea pigs. *Resp. Physiol.* **11**, 135.
28. Perutz, M. F. (1970). Stereochemistry of cooperative effects in haemoglobin. *Nature (London)* **228**, 726.
29. Schmidt-Nielsen, K., and Larimer, J L. (1958). Oxygen dissociation curves of mammalian blood in relation to body size. *Amer. J. Physiol.* **195**, 424.
30. Chiodi, H. (1970-1971). Comparative study of the blood gas transport in high altitude and sea level Camelidae and goats. *Resp. Physiol.* **11**, 84.
31. Hall, F. G. (1965). Hemoglobin and oxygen affinities in seven species of Sciuridae. *Science* **148**, 1350.
32. Bauer, C., Ludwig, M., Ludwig, I., and Bartels, H. (1969). Factors governing the oxygen affinity of human adult and foetal blood. *Resp. Physiol.* **7**, 271.
33. Barker, J. N. (1957). Role of hemoglobin affinity and concentration in determining hypoxia tolerance of mammals during infancy, hypoxia, hyperoxia and irradiation. *Amer. J. Physiol.* **189**, 281.
34. Chiodi, H. (1962). Oxygen affinity of the hemoglobin of high altitude mammals. *Acta Physiol. Lat. Amer.* **12**, 208.
35. Rahn, H. (1966). Introduction to the study of man at high altitudes: Conductance of O_2 from the environment to the tissues. *In* "Life at High Altitudes," p. 2. Pan Amer. Health Organ., Washington, D.C.
36. Valdivia, E. (1958). Total capillary bed in striated muscle of guinea pigs native to the Peruvian mountains. *Amer. J. Physiol.* **194**, 585.
37. Reynafarje, C. (1966). Physiological patterns: Hematological aspects. *In* "Life at High

Altitudes," p. 40. Pan Amer. Health Organ., Washington, D.C.

38. McGrath, J. J., and Bullard, R. W. (1968). Altered myocardial performance in response to anoxia after high-altitude exposure. *J. Appl. Physiol.* **25**, 761.
39. Ou, L. C., and Tenney, S. M. (1970). Properties of mitochondria from hearts of cattle acclimatized to high altitude. *Resp. Physiol.* 8, 151.
40. Burlington, R. F., and Maher, J. T. (1968). Effect of anoxia on mechanical performance of isolated atria from ground squirrels and rats acclimatized to altitude. *Nature (London)* **219**, 1370.
41. Komives, G. K., and Bullard, R. W. (1967). Function of the phrenic nerve-diaphragm preparation in acclimation to hypoxia. *Amer. J. Physiol.* **212**, 788.
42. Magez, M., Blatt, W. F., Natele, P. J. and Blatteis, C. M. (1968). Effect of high altitude on lactic dehydrogenase isozyme of neonatal and adult rats. *Amer. J. Physiol.* **215**, 8.
43. Souhrada, J., and Bullard, R. W. (1971). The influence of pH on contractile function of the hamster myocardium. *Cryobiology* **8**, 159.

XVI

Body Fluids, Body Composition, and Metabolic Aspects of High-Altitude Adaptation

C. F. CONSOLAZIO, H. L. JOHNSON, and H. J. KRZYWICKI

Introduction

The marked incidence and severity of acute mountain sickness (AMS) resulting from abrupt altitude exposure is a major problem in maintaining maximal efficiency of troops and could seriously impair military performance and operations. There is lack of agreement as to the extent of this reduction, its underlying causes, its operational implications, and the most practical means of alleviating AMS. This information is a necessary requirement for an accurate estimate of the operational capacity of combat troops.

It has been consistently shown that abrupt altitude exposure results in anorexia. This loss of appetite results in a caloric deficit, with a subsequent body weight loss. The question has arisen as to whether some of the abnormal metabolic or biochemical changes observed at altitude may be due to caloric restriction rather than hypoxia. For example, during acute altitude exposure, one observes negative nitrogen and water balances (1, 2), decreased fasting glucose levels and lowered glucose disappearance curves (3), and abnormal changes in electrolyte metabolism (2, 4). These abnormalities are similar to those observed during acute starvation and caloric restriction (5, 6).

In this chapter, I would like to discuss some of the altitude studies that we have conducted at Pike's Peak, Colorado (4300 m) during the past 6 years (Tables Ia, Ib, and Ic).

In our first Pike's Peak study in 1964, the acute mountain sickness (AMS)

TABLE Ia. Experimental Design 1964 (4300 m)[a]

To evaluate the effects of gradual to abrupt altitude exposure in sea level natives (two groups of eight men each)

a. Work performance (maximal and submaximal)
b. Mountain sickness symptoms
c. Water and nitrogen balances
d. Arterial acid-base blood parameters
e. Psychological evaluation

[a] All men had severe AMS at approximately 36 hr, but by 72 hr the symptoms had greatly decreased. Food intakes were reduced in all during altitude exposure.

TABLE Ib. Altitude 1967 (4300 m)[a]

To evaluate the effects of normal and high carbohydrate diets during abrupt altitude exposure (sea level natives)

a. Acute mountain sickness symptoms
b. Physical conditioning prior to altitude exposure
c. Nitrogen, water, and mineral metabolism
d. Body compartments
e. Various levels of submaximal work performance (treadmill)

	Diet, all liquid	
	Normal nutrients	*High carbohydrate*
Physically conditioned	5	9
Food intake	Decreased	Decreased

[a] All men had greatly reduced AMS, with maximal severity at 6-10 hr of acute exposure.

symptoms were greatly decreased in the group ascending to altitude gradually (7) The group exposed abruptly had their maximum severity of AMS at 36 hr, and by 72 hr most of the severe symptoms had disappeared. Although this was an interesting observation, our main interest continued in the area of abrupt altitude exposure and the soldier who had to be physically and mentally alert during acute exposure.

Food intakes were decreased by 8.2% and 10.0% during the 28-day altitude exposure, for the gradual and abrupt groups, respectively; although nitrogen intakes were adequate, nitrogen balances were negative showing a cumulative loss of 19.0 and 67.2 gm for the 28-day period (1). This was the first in a series of studies in which we observed negative water balances at altitude, under conditions where fluids were available *ad libitum* (Table II). These losses averaged 2.69 and 1.88 kg for groups I and II, respectively.

The observations from this study led to the following conclusions: (a) gradual

TABLE Ic. Altitude 1968 (4300 m)

To again evaluate the effects of normal and high carbohydrate diets on men eating normally at altitude (eight men in each group)

a. Acute mountain sickness symptoms
b. Physical conditioning versus no conditioning
c. Fluid compartments and balances
d. Nitrogen and mineral balances, blood electrolyte levels
e. Fasting glucose and glucose tolerance curves
f. Submaximal and maximal work performance (treadmill)

	Diet	
	Normal	*High carbohydrate*
Physically conditioned	6 men[a]	6 men[a]
No conditioning (sedentary)	2 men[b]	2 men[b]

[a]These men consumed between 3400-3550 Kcal/day at altitude. In all of them the severity AMS was greatly reduced.
[b]Severe AMS, with anorexia, etc. These men were hospitalized at 20 and 40 hr of acute altitude exposure.

TABLE II. High Altitude 1964 (4300 m)[a]

	Groups	
Changes (28 days)	*Gradual*	*Abrupt exposure*
Body weight (kg)	– 2.66	– 3.80
Nitrogen balance (gm)	–19.00	–67.20
Caloric deficit (Kcal)	–5404 (–8.2%)	–6048 (–10.0%)
Caloric deficit, Body wt equivalent (kg)	0.70	0.79
Water balance (kg)	– 2.69	– 1.88

[a]Mean of eight men in each group – normal military ration.

exposure to altitude was beneficial in reducing the AMS symptoms; (b) the negative nitrogen balances appeared to be due to decreased utilization of protein and the possibility of a decrease in protein synthesis; and (c) the negative water balances were difficult to explain at the time, although hyperventilation and the low humidity were factors.

Although other studies were done in 1965 and 1966, one significant study was completed in the summer of 1967 (8, 9). This study was the result of an excellent literature review by Mitchell and Edman (10) on the effects of high carbohydrate intakes upon altitude tolerance. In their summary, it appeared that high carbohydrate diets were beneficial in (a) increasing the work capacity, (b) reducing AMS severity, (c) delaying syncope for longer periods, and (d) increasing mental efficiency.

Our study (8) was designed to investigate some of these problems and to evaluate (a) the body fluids and body compartment changes, (b) the nitrogen and mineral balances, and (c) the effects of heavy physical conditioning prior to acute altitude exposure. Two groups were studied. Group I received a diet containing a normal distribution of nutrients, and group II received a diet containing high carbohydrate (68% of the calories).

Although anorexia occurred, resulting in a 22-33% decrease in food intake at altitude, the data indicated that maintaining heavy physical activity and conditioning greatly reduced the severity of AMS and that the high carbohydrate group showed an even greater reduction (8) (Table III).

These observations led to further studies on the effects of high carbohydrate diets and altitude exposure during the summer of 1968. This study also had a normal and high carbohydrate group and was designed to implement some of the findings of the previous study under conditions in which the men were eating normally at high altitude. To induce greater consumption, a combination of normal foods and liquid diets were used. Daily intakes at altitude were fairly high, averaging 3524 and 3416 Kcal/day for the normal and high carbohydrate groups, respectively (Table IV and Fig. 1). The 1967 study, in which anorexia occurred, and the 1968 study, in which the men consumed a normal intake at altitude will be compared (8).

Acute Mountain Sickness Symptoms

The AMS symptoms were drastically reduced in both studies during acute altitude exposure and appeared to have disappeared after the first evening at altitude. The short duration of these severe symptoms should be emphasized, since earlier studies (7) have shown that the severity of clinical mountain sickness symptoms increased and reached a maximum between 30-36 hr of altitude exposure.

None of the physically conditioned men, whether they consumed a normal or a high carbohydrate diet, were seriously sick at altitude. The maximal severity of AMS was observed between 6 to 10 hr of acute exposure, and, by morning, all of the men appeared to be in fairly good health. The same was also true for the sedentary group consuming a high carbohydrate diet. These men all ate normally at altitude, indicating the beneficial effects of physical conditioning and a normal food intake in reducing the severity of AMS.

However, a sedentary existence, with a consumption of a diet containing a normal percentage of nutrients, resulted in near disaster for these two men

TABLE III. 1967 Altitude Study[a]

	Summary			
	Regular diet		High carbohydrate diet	
Parameter	control	altitude	control	altitude
Body wt changes (kg)		−3.81		−4.80
Caloric intake/day	3100	2015.00	3179	1915.00
Protein intake (day/gm)	104.1	62.10	104.6	51.10
Nitrogen balance (gm/day)	+2.68	−1.16	+2.98	−2.44
Water balance (kg/period)		−2.11		−2.52

[a]Means for a 12-day period at 4300 m.

(Table Ic). Both men were violently sick and suffered from severe headaches, nausea, vomiting, diarrhea, etc. One individual had to be removed from altitude at 20 hr, and the second, who appeared to have a delayed reaction, was removed at 40 hr of exposure and hospitalized. Both required electrolyte therapy for 2 days before being released.

Body Weight

The major difference between 1967 (8) and the 1968 studies was anorexia and the subsequent large body weight losses observed at altitude in 1967. In both studies, the men had exercised strenuously prior to altitude exposure, which may have been beneficial in decreasing the severity of the clinical symptoms at altitude. In all of our studies at altitude, we find that it may be beneficial to keep subjects outdoors and exercising moderately during the day. With moderate exercise, they soon appear to learn to hyperventilate more rapidly. This is contrary to the thinking of some investigators who feel that exercise at altitude should be greatly reduced during the first few days of exposure.

Larger body weight losses accompanied the anorexia in the 1967 study. The 6-day altitude phase showed an average body weight loss of 2.09 and 2.65 kg for the normal group I, and high carbohydrate group II, respectively. In the 1968 study, in which the men ate normally at altitude, the body weight losses were smaller for the same period, averaging 0.88 and 1.13 kg for the same respective groups (Table V). These data indicated that some of the body weight losses could be attributed to the caloric deficit in the 1967 study, but some losses not attributable to reduced caloric intakes also occurred in both studies. These

TABLE IV. Altitude (4300 m). Comparison of 1967-1968 Studies – Calories Consumed/Day[a]

	1967		1968	
Phase	*I*	*II*	*I*	*II*
Control	3070	3160	3802	3639
Altitude				
Days 1-3	2493	2260	3511	3513
4-6	2323	1950	3537	3319
Consumed at altitude (%)	78.4	66.6	92.7	93.9
Rehabilitation				
Days 1-6	2680	2540	3747	3460

[a]Group I, normal diet; group II, high carbohydrate diet. The 1967 study was calculated on a 6-day basis to be comparable to the 1968 study.

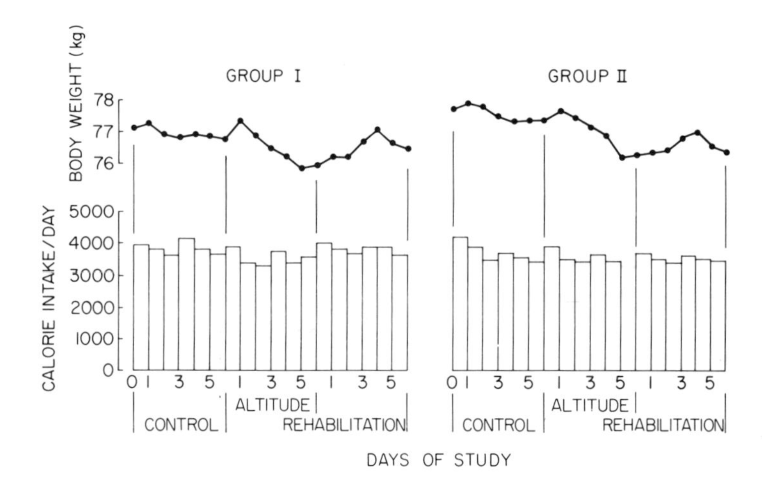

FIG. 1. Effects of high altitude on body weight and caloric intake in active subjects.

unexplained weight losses are probably the result of hypohydration during early altitude exposure, which will be discussed later in the text.

Nitrogen Metabolism

Nitrogen balances, exclusive of sweat losses, were again negative in the 1967 study, averaging a loss of 0.11 and 1.90 gm/day for groups I and II, respectively.

TABLE V. Altitude (4300 m). Comparison of 1967-1968 Studies – Body Weight Changes (kg)[a]

	1967		*1968*	
Phase	*I*	*II*	*I*	*II*
Control	70.04	75.25	76.69	77.10
Altitude				
Days 1-3	69.10	74.08	76.42	76.84
4-6	67.95	72.60	75.81	75.97
Body weight loss at altitude	−2.09	−2.65	−0.88	−1.13

[a]Group I, normal diet; group II, high carbohydrate diet. The 1967 was study calculated on a 6-day basis to be comparable to the 1968 study.

However, in the 1968 study, nitrogen balances, exclusive of sweat losses, were highly positive in all phases, averaging a +11.38 and +11.97 gm/day for groups I and II during acute altitude exposure (Table VI). Negative nitrogen balances during short term altitude exposure with daily intakes of 50-60 gm of protein have been reported (1, 11). This has led to questions regarding the adequacy of the NRC's daily allowances (12) of 1 gm of protein/kg body weight during altitude exposure; these allowances appear to be more than adequate in maintaining nitrogen balances under normal conditions. Surks (11) observed negative nitrogen balances with intakes of 0.9 gm of protein/kg body weight; and in serum albumin turnover studies, they indicated that altitude exposure could have caused some impairment in protein biosynthesis during the first week of altitude exposure. With adequate calorie and protein intakes, as in the 1968 study, nitrogen retention was observed for the first time during altitude exposure. Rather than the protein losses of 4 and 71 gm as noted in the 1967 study, protein retentions of 430 and 448 gm/man were observed during the 6-day altitude phase. Therefore, it appears that the protein catabolism observed in earlier studies, concomitant with 30-50% reductions in protein and calorie consumption, may be totally attributable to anorexia rather than a direct effect of hypoxia.

Mineral Metabolism

In the 1968 study, the serum sodium, potassium, and calcium levels were practically unchanged during high-altitude exposure. Some investigators have reported no changes in serum potassium levels (4, 13); some have reported

TABLE VI. Altitude (4300 m) Comparison of 1967-1968 Studies – Nitrogen Balance (gm/day)[a]

	1967		*1968*	
Phase	*I*	*II*	*I*	*II*
Control			+ 8.47	+10.59
Altitude				
Days 1-3	–0.32	–1.45	+13.10	+13.00
4-6	+0.11	–2.34	+ 9.67	+10.87
Mean/day at altitude	–0.11	–1.90	+11.38	+11.94
Rehabilitation				
Days 1-3	+1.95	–0.03	+ 6.36	+ 8.63
4-6	+2.50	+1.05	+ 7.30	+ 8.58

[a]Group I, normal diet; group II, high carbohydrate diet. Does not include sweat losses.

decreased potassium (14-16); and others have reported an increase in potassium levels during altitude exposure (17). As expected, this has led to a very confusing picture regarding the relationship of serum electrolyte levels and altitude exposure.

In other studies (2, 4), the daily urinary excretions of sodium were essentially unchanged, but potassium excretions were decreased during altitude exposure. In these studies (2, 4), the daily food intakes were drastically reduced at altitude, which may have accounted for these changes. In the 1968 study, although decreased urinary potassium excretions were observed during days 1-3 of altitude exposure, by day 4 these values were comparable to control values. However, potassium balances at altitude were essentially positive in both the 1967 and 1968 studies. Sodium balances, exclusive of sweat losses, were positive throughout the study and were not affected by the differences in the two diets (Table VII). Although the positive sodium balances may be indicative of some water retention, one must keep in mind that these balances did not include the daily sodium losses observed in sweat. Past studies have indicated that daily sodium losses in sweat can be fairly high (2.8-6.0 gm) even in acclimated subjects under conditions of profuse sweating (18). It appears that a large portion of the positive sodium balance, observed in the 1968 study, could be attributed to the fairly high sweat losses. Since other complete electrolyte balance studies in humans at altitude have not been reported, it has not been possible to compare our reported findings with other data.

As a final word on electrolyte metabolism, it appears that the urinary electrolyte excretions and serum levels, alone, do not provide sufficient information to thoroughly evaluate mineral metabolism in humans, and one

TABLE VII. Altitude (4300 m). Comparison of 1967-1968 Studies – Mineral Balances (gm/day)[a]

Phase	*1967 (Anorexia)*		*1968 (Normal)*	
	Sodium	*Potassium*	*Sodium*	*Potassium*
Control	+0.05	–0.10	+7.05	+0.46
Altitude				
Days 1-3	–1.59	–0.39	+6.70	+1.67
4-6	–1.29	0	+5.15	+0.29
Rehabilitation				
Days 1-3	+0.10	–0.20	+6.20	–0.72
4-6	–0.21	+0.05	+5.92	–0.40

[a]Since there were no significant differences between the groups, they were combined for comparisons.

must also evaluate the daily intakes of these nutrients to draw any definite conclusions. It is indicated that the normal consumption of food, including the essential nutrients, may be very beneficial in maintaining electrolyte balances during altitude exposure.

Blood Sugar Metabolism

Varied and inconsistent changes have been reported in fasting blood sugar levels and glucose tolerance curves at high altitude. Forbes (19) studied three subjects at 5300 m and observed that glucose curves were increased in two men who had been at this altitude for 17 days and lowered in the third man who had been at altitude for only 6 days. Subsequent studies by Picon-Reategui (20) utilizing both oral and intravenous glucose tolerance tests revealed a faster rate of glucose utilization in high altitude residents as compared with those at sea level. However, none of these studies discussed these changes in relation to the daily food intakes. More recently, Janoski *et al.* (3) performed both oral and intravenous glucose tolerance tests on 10 subjects at sea level and at 4300 m and showed that at altitude the glucose disappearance was lower than the respective values at sea level. However, in this study (3) the daily caloric intakes were very low, averaging 900 Kcal/man for a 7-day period at altitude. The question arises as to whether these changes in glucose disappearance were the results of hypoxia or of the great caloric restriction during altitude exposure. In the 1968 study, it was observed that fasting blood glucose levels and glucose tolerance curves were normal at high altitude, which indicated that when food intakes are normal at altitude, glucose utilization appears to be normal (Fig. 2).

Body Water Metabolism

Negative water balances were again observed in both studies, even though adequate nutrient intakes, including minerals and water, were consumed in the 1968 study (Table VIII).

During the first 6 days of the altitude phase, the two groups averaged losses of 1791 and 2241 gm of water with reduced caloric intakes, while these losses averaged 1683 and 753 gm in the 1968 study despite adequate caloric consumption. Urine volumes and specific gravities indicated that fluid consumption was adequate, so this was not voluntary water deprivation. Water retention was observed in all groups after their return to sea level. This is contrary to the data and conclusions of Hannon *et al.* (21), who advocate a redistribution of body fluids between fluid spaces at altitude. To resolve these discrepancies, body water compartments were measured in the 1967-1968 studies.

In the 1967 study, comparisons of the body compartment changes were made during the 12-day altitude phase (Table IX). During altitude exposure, body weights were decreased, averaging a –3.54 and –3.96 kg for the control and high carbohydrate groups, respectively. Body fat accounted for a –1.29 and –1.46 kg, and body proteins were also decreased, averaging a –0.32 and –0.47 kg, and body minerals were decreased by 0.16 and 0.18 kg, respectively. Total body water losses were fairly large, showing a loss of 1.77 and 1.85 kg for the same respective groups (9). The total body water losses and the water balance data were comparable.

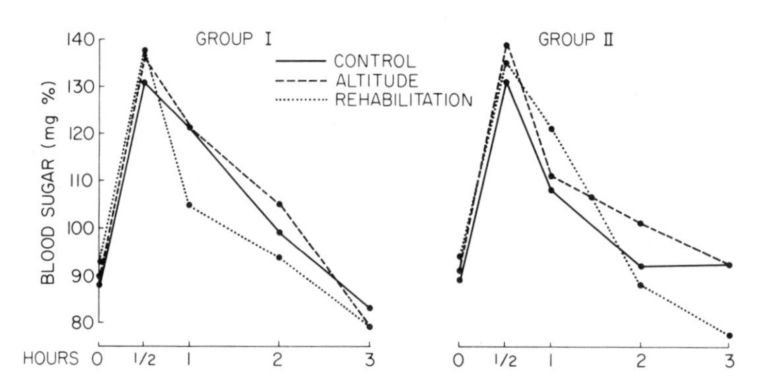

FIG. 2. Effects of high altitude on glucose tolerance curves. There were no significant differences either between the two groups or between phases.

TABLE VIII. Altitude (4300 m). Comparison of 1967-1968 Studies – Water balances (gm/day)[a]

	1967		*1968*	
Phase	*I*	*II*	*I*	*II*
Control	+ 29	−44	+ 12	− 98
Altitude				
Days 1-3	−353	−565	−170	−137
4-6	−244	−182	−391	−114
Total change at altitude	−1791	−2241	−1683	−753
Rehabilitation				
Days 1-3	+275	+259	+209	+192
4-6	− 11	+201	−506	−122

[a]Comparisons for 6-day periods.

TABLE IX. High Altitude 1967 (4300 meters) – Body Fluids and Body Composition[a]

Body compartments	*Group I (normal)*	*Group II (high carbohydrate)*
n (kg)		
Body weights	−3.54	−3.96
Body fat	−1.29	−1.46
Body protein	−0.32	−0.47
Body minerals	−0.16	−0.18
Body water	−1.77	−1.85
(ml)		
Blood volume	−884 (16%)[b]	−969 (15.6%)[b]
Plasma volume	−716 (22.5%)[b]	−760 (21.3%)[b]
RBC volume	−168	−209

[a]Changes during 12 days of altitude exposure (14 men). [b]Significantly different from control values.

Other important factors were the blood and plasma volume data at altitude. Blood volumes were significantly decreased by 880 gm (16%) and 970 gm (15.6%) during acute altitude exposure. The same was true for the plasma volume data, which was decreased by 22.5% and 21.3% for groups I and II during altitude exposure (Table IX). In the 1968 study, both groups again showed a decrease in total body water at 6 days of altitude exposure (Table X). Since no significant differences were observed between groups, they were combined. The combined data showed a significant loss of 2.25 kg total body water after 6 days of altitude exposure. Although extracellular water values (determined by thiocyanate dilution), were increased at altitude, they were not

TABLE X. Altitude 1968 (4300 m) – Water Compartments

Water Content (kg)	*Sea level*	*Altitude*	*Rehabilitation sea level*
Total body water (D_2O dilution)	42.00	39.75[a]	39.72
Extracellular water (thiocyanate)	19.48	20.75	21.29
Intracellular water	22.52	19.00[a]	18.43

[a]Significantly different from controls.

significantly different from control values. The calculation of intracellular waters, by difference, showed them to be decreased by 2.52 kg during altitude exposure.

The effects of altitude upon body water metabolism has been a controversial subject among investigators. Surks *et al.* (22) and Hannon *et al.* (21) have indicated that the decrease in blood volumes, particularly plasma, at altitude may result from a redistribution of body fluids rather than from hypohydration. On the other hand, several investigators (1, 9) have consistently shown body hypohydration during altitude exposure by observing large decreases in total body water, plasma and blood volumes, and negative water balances. In all instances, the fairly large body weight losses occurring during altitude exposure could not all be accounted for by the caloric deficit.

In the 1967 study, it appeared that the weight loss was partially due to the caloric deficit and the remainder as a loss in total body water, but, in the 1968 study, the body weight loss was primarily due to total body water loss since there was practically no caloric deficit during altitude exposure. Body water losses (from D_2O dilution) were greater than weight losses, but this can be accounted for by an increase in body protein because nitrogen balances were positive. Hannon *et al.* (21) has stressed the shift in body fluids from extracellular to intracellular water during acute altitude exposure, and Carson *et al.* (23) has reported that the maximal severity of mountain sickness symptoms occurs during the period of the greatest fluids shifts. However, they could not account for their body weight changes from caloric deficit alone, and thus a loss from some other body compartment was indicated.

Malpartida and Moncloa (24) recently showed the radiosulfate space to be reduced by 1.4 liters during 48 hr of acute altitude exposure. It is unfortunate that water balances were not reported in any of these studies. Picon-Reategui *et al.* (25) reported that 94% of the total body weight loss in rats at simulated altitudes of 15,000 ft was due to body water loss. Asmussen and Nielsen (26) reported a hypoxia diuresis at altitude. Recently, Singh *et al.* (27) indicated that furosemide, a diuretic, had been effectively used to reduce the AMS in Indian troops, but it would appear that this would be harmful and impractical since the diuretic would impose an extra water loss in individuals who are already hypohydrated. This is based on information from our 1967 and 1968 studies, in which we observed a natural "cold" diuresis during the first two days of altitude exposure (averaging an increase of 1107 gm in urine volumes). Under these conditions, the troops showed a great reduction in AMS severity and were in good physical condition. Our data strongly indicated that these water losses were at least partially due to the altitude exposure and may be an adaptive response of the body to the hypoxic environment.

It appears that on arrival to altitude, the men should not be permitted to lie down and go to sleep. We recommend that they be kept outdoors and exercise moderately, which will allow them to learn to hyperventilate more rapidly. In addition, the fairly cold altitude environment usually results in a natural cold diuresis, which appears to be very beneficial in reducing the severity of AMS symptoms.

Summary

Anorexia accompanied by body weight losses and negative water and nitrogen balances were observed in the 1964 and 1967 studies. In the 1967 and 1968 studies, AMS was drastically reduced. The daily caloric consumption in human subjects can be maintained after abrupt exposure to high altitude, providing the men have reduced severity of AMS and are in good physical condition prior to and during altitude exposure. This study indicates that (a) positive nitrogen balances can be achieved during altitude exposure; (b) body weight losses can be greatly reduced; (c) mineral balances are positive; (d) blood electrolyte levels are normal; and (e) fasting glucose levels and glucose tolerance curves are normal. Normal digestion and absorption occurred during acute altitude exposure. It appears that many of the biochemical changes that occur during abrupt high altitude exposure, previously attributed to hypoxia, may be the result of anorexia and the subsequent caloric restriction.

Since total body water and intracellular water are significantly decreased during acute altitude exposure, it appears that hypohydration and a natural diuresis occurs. This appears to be an adaptive mechanism that may reduce cerebrospinal pressure, and cerebral edema, the presence of which reduces the severity of AMS. The severity of AMS during abrupt altitude exposure can be greatly reduced without the use of drug therapy. The following conditions appear to be highly beneficial: (a) heavy physical activity and physical conditioning prior to altitude exposure; (b) the consumption of a minimal quantity of carbohydrate (at least 320 gm/day), (c) the maintenance of a normal food intake at altitude, and (d) a natural "cold" diuresis, with a subsequent decrease in total body water and intracellular water.

References

1. Consolazio, C. F., Matoush, L. O., Johnson, H. L. and Daws, T. A. (1968). Protein and water balances in young adults during prolonged exposure to high altitude (4300 m). *Amer. J. Clin. Nutr.* **21**, 154.
2. Johnson, H. L., Consolazio, C. F., Matoush, L. O., and Krzywicki, H. J. (1969). Nitrogen and mineral metabolism at altitude. *Fed. Proc., Fed. Amer. Soc. Exp. Biol.* **28**, 1195.
3. Janoski, A. H., Johnson, H. L., and Sanbar, S. S. (1969). Carbohydrate metabolism in man at altitude. *Fed. Proc., Fed. Amer. Soc. Exp. Biol.* **28**, 593.
4. Janoski, A. H., Whitten, B. K., Shields, J. L., and Hannon, J. P. (1969). Electrolyte patterns and regulation on man during acute exposure to high altitude. *Fed. Proc., Fed. Amer. Soc. Exp. Biol.* **28**, 1185.
5. Consolazio, C. F., Matoush, L. O., Johnson, H. L., Nelson, R. A., and Krzywicki, H. J. (1967). Metabolic aspects of acute starvation in normal humans (10 days). *Amer. J. Clin. Nutr.* **20**, 672.
6. Krzywicki, H. J., Consolazio, C. F., Matoush, L. O., and Johnson, H. L. (1968). Metabolic aspects of acute starvation, body composition changes. *Amer. J. Clin. Nutr.* **21**, 87.
7. Evans, W. L. (1966). Measurement of subjective symptomatology of acute high altitude sickness. *Psychol. Rep.* **19**, 815.
8. Consolazio, C. F., Matoush, L. O., Johnson, H. L., Krzywicki, H. J., Daws, T. A., and Isaac, G. J. (1969). Effects of high-carbohydrate diets on performance and clinical symptomatology after rapid ascent to high altitude. *Fed. Proc., Fed. Amer. Soc. Exp. Biol.* **28**, 937.
9. Krzywicki, H. J., Consolazio, C. F., Matoush, L. O., Johnson, H. L., and Barnhart, R. A. (1969). Body composition changes during exposure to altitude. *Fed. Proc., Fed. Amer. Soc. Exp. Biol.* **28**, 1190.
10. Mitchell, H. H., and Edman, M. (1949). "Nutrition and Resistance to Climatic Stress." Office of the Quartermaster General (OQMG), QMF and CI, Chicago, Illinois.
11. Surks, M. I. (1966). Metabolism of human serum albumin in man during acute exposure to high altitude (14,100 ft). *J. Clin. Invest.* **45**, 1442.
12. National Academy of Science–National Research Council (1968). "Recommended Dietary Allowances," 7th rev. ed., Publ. No. 1694. Nat. Acad. Sci.–Nat. Res. Counc., Washington, D.C.
13. Dill, D. B., Talbott, J. H., and Consolazio, W. V. (1937). Blood as a physiochemical system. XII. Man at altitude. *J. Biol. Chem.* **118**, 649.
14. Elkinton, J. R., Singer, R. B., Barker, E. S., and Clark, J. K. (1955). Effects in man of acute environmental respiratory alkalosis and acidosis on ionic transfers in the total body fluids. *J. Clin. Invest.* **34**, 1671.
15. Ferguson, F. P., Smith, D. C., and Barry, J. Q. (1957). Hypokalemia in adrenalectomized dogs during acute decompression stress. *Endocrinology* **60**, 761.
16. Gold, A. J., Barry, J. Q., and Ferguson, F. P. (1961). Relation of respiratory alkalosis to

hypokalemia in anesthetized dogs during altitude stress. *J. Appl. Physiol.* **16**, 837.
17. Jezek, V., Ouredvik, A., Daum, S., and Krouzkoa, L. (1965). Effects of short term hypoxemia on the sodium and potassium metabolism and on cardiac contraction. *Acta Med. Scand.* **177**, 175.
18. Consolazio, C. F., Matoush, L. O., Nelson, R. A., Harding, R. S., and Canham, J. E. (1963). Excretion of sodium, potassium, magnesium and iron in human sweat and the relation of each to balance and requirements. *J. Nutr.* **79**, 407.
19. Forbes, W. H. (1936). Blood sugar and glucose tolerance at altitude. *Amer. J. Physiol.* **116**, 309.
20. Picon-Reategui, E. (1962). Studies on the metabolism of carbohydrates at sea level and at high altitude. *Metab. Clin. Exp.* **11**, 1148.
21. Hannon, J. P., Chinn, K. S. K., and Shields, J. L. (1969). Effects of acute high altitude exposure on body fluids. *Fed. Proc., Fed. Amer. Soc. Exp. Biol.* **28**, 1178.
22. Surks, M. I., Chinn, K. S. K., and Matoush, L. O. (1966). Alterations in body composition in man after acute exposure to high altitude. *J. Appl. Physiol.* **21**, 1741.
23. Carson, R. P., Evans, W. O., Shields, J. L., and Hannon, J. P. (1969). Symptomology, pathophysiology and treatment of acute mountain sickness. *Fed. Proc., Fed. Amer. Soc. Exp. Biol.* **28**, 1085.
24. Malpartida, M., and Moncloa, F. (1967). Radiosulfate space in humans at high altitude. *Proc. Soc. Exp. Biol. Med.* **125**, 1328.
25. Picon-Reategui, E., Fryers, G. R., and Berlin, N. I. (1953). Effect of reducing the atmospheric pressure in body water content of rats. *Amer. J. Physiol.* **172**, 33.
26. Asmussen, E., and Nielsen, M. (1945). Studies on the initial increase in O_2 capacity of the blood at low O_2 pressure. *Acta Physiol. Scand.* **9**, 75.
27. Singh, I., Khanna, P. K., Srevosta, M. C., and Subramanyam, C. S. V. (1969). Acute mountain sickness. *N. Eng. J. Med.* **280**, 175.

Author Index

Numbers in parentheses are reference numbers and indicate that an authors work is referred to, although his name is not cited in the text. Numbers in italics show the page on which the complete reference is listed.

I

J

K

Subject Index